GRADES 7-8

the Super Source™
Patterns and Functions

Cuisenaire Company of America, Inc.
White Plains, NY

Cuisenaire extends its warmest thanks to the many teachers and students across the country who helped ensure the success of the Super Source™ series by participating in the outlining, writing, and field testing of the materials.

Managing Editor: Alan MacDonell
Developmental Editor: Deborah J. Slade
Contributing Writer: Carol Desoe
Production/Manufacturing Director: Janet Yearian
Production Coordinator: Joan Lee

Design Director: Phyllis Aycock
Cover Design: Phyllis Aycock
Line Art and Composition: Eileen Sullivan

Copyright © 1998 by
Cuisenaire Company of America, Inc.
10 Bank Street, White Plains, New York 10602

All rights reserved
Printed in the United States of America
ISBN 1-57452-173-X
CCA015187

Permission is granted for limited reproduction of pages from this book for classroom use.
The word CUISENAIRE and the color sequence of the rods, cubes, and squares are trademarks of Cuisenaire Company of America, Inc., registered in the United States and other countries.

1 2 3 4 5 - SG - 02 01 00 99 98

the Super Source™
Table of Contents

INTRODUCTION 4
OVERVIEW OF THE LESSONS 6

LESSONS

Investigating Growth Patterns 8
   Backyard Improvements 9
   Ripples 14
   Marquetry 19
   The Pyramid Mystery 24
   Bees in the Trees 29

Investigating Patterns from Geometry 34
   Napkins and Place Mats 35
   Pythagoras Delivers the Mail 40
   Polygons, Pegs, and Patterns 44
   The Airline Connection 48
   Inside Out, Outside In 52

Investigating Patterns and Functions as Problem-Solving Tools 57
   Count Square and Countess Triangle 58
   Greek Border Designs 63
   Table for 63, Please 68
   Birthday Cakes 73

Investigating Other Topics Using Patterns and Functions 77
   Visual Effects 78
   Pascal Pastimes 83
   Beehive Buzz 88
   Carol's Kite Kits 93

BLACKLINE MASTERS
   Activity Masters 99
   Geodot Paper 117
   Circular Geodot Paper 118
   Dot Paper 119
   Color Tile Grid Paper 120
   1-centimeter Grid Paper 121
   Pattern Block Triangle Paper 122
   Isometric Dot Paper 123

Using the Super Source™

The Super Source™ Grades 7-8 continues the Grades K-6 series of activities using manipulatives. Driving **the Super Source** is Cuisenaire's conviction that children construct their own understandings through rich, hands-on mathematical experiences. There is a substantial history of manipulative use in the primary grades, but it is only in the past ten years that educators have come to agree that manipulatives play an important part in middle grade learning as well.

Unlike the K-6 series, in which each book is dedicated to a particular manipulative, the Grades 7-8 series is organized according to five curriculum strands: Number, Geometry, Measurement, Patterns/Functions, and Probability/Statistics. The series includes a separate book for each strand, each book containing activities in which students use a variety of manipulatives: Pattern Blocks, Geoboards, Cuisenaire® Rods, Snap™ Cubes, Color Tiles, and Tangrams.

Each book contains eighteen lessons grouped into clusters of 3-5 lessons each. Each cluster of lessons is introduced by a page of comments about how and when the activities within each lesson might be integrated into the curriculum.

The lessons in **the Super Source** follow a basic structure consistent with the vision of mathematics teaching described in the *Curriculum and Evaluation Standards for School Mathematics* published by the National Council of Teachers of Mathematics. All of the activities involve Problem Solving, Communication, Reasoning, and Mathematical Connections—the first four NCTM Standards.

HOW THE LESSONS ARE ORGANIZED

At the beginning of each lesson, you will find, to the right of the title, a list of the topics that students will be working with. GETTING READY, offers an *Overview*, which states what children will be doing, and why, and provides a list of *What You'll Need*. Specific numbers of manipulative pieces are suggested on this list but can be adjusted as the needs of your particular situation dictate. In preparation for an activity, pieces can be counted out and placed in containers or self-sealing plastic bags for easy distribution. Blackline masters that are provided for your convenience at the back of the book are also referenced on this materials list, as are activity masters for each lesson.

The second section, THE ACTIVITY, contains the heart of the lesson: a two-part *On Their Own*, in which rich problems stimulate many different problem-solving approaches and lead to a variety of solutions. These hands-on explorations have the potential for bringing students to new mathematical ideas and deepening skills. They are intended as stand-alone activities for students to explore with a partner or in a small group. *On Their Own* Part 2 extends the mathematical ideas explored in Part 1.

After each part of *On Their Own* is a *Thinking and Sharing* section that includes prompts you can use to promote discussion. These are questions that encourage students to describe what they notice, tell how they found their results, and give the reasoning behind their conclusions. In this way, students learn to verify their own results rather than relying on the teacher to determine if an answer is "right" or "wrong." When students compare and discuss results, they are often able to refine their thinking and confirm ideas that were only speculations during their work on the *On Their Own* activities.

The last part of THE ACTIVITY is *For Their Portfolio*, an opportunity for the individual student to put together what he or she has learned from the activity and discussion. This might be a piece of writing in which the student communicates results to a third person; it could be a drawing or plan synthesizing what has occurred; or it could be a paragraph in which the student applies the ideas from the activity to another area. In any case, the work students produce *For Their Portfolio* is a reflection of what they've taken from the activity and made their own.

The third and final section of the lesson is TEACHER TALK. Here, in *Where's the Mathematics?*, you as the teacher can gain insight into the mathematics underlying the activity and discover some of the strategies students are apt to use as they work. Solutions are also given, when such are necessary and/or helpful. This section will be especially helpful to you in focusing the *Thinking and Sharing* discussion.

USING THE ACTIVITIES

The Super Source is designed to fit into a variety of classroom environments. These can range from a completely manipulative-based classroom to one in which manipulatives are just beginning to play a part. You may choose to have the whole class work on one particular activity, or you may want to set different groups to work on two or three related activities. This latter approach does not require full classroom sets of a particular manipulative.

If students are comfortable working independently, you might want to set up a "menu"—that is, set out a number of activities from which students can choose. If this is the route you take, you may find it easiest to group the lessons as they are organized in the book—in small clusters of related activities that stimulate similar questions and discussion.

However you choose to use **the Super Source** activities, it would be wise to allow time for several groups or the entire class to share their experiences. The dynamics of this type of interaction, where students share not only solutions and strategies but also thoughts and intuitions, is the basis of continued mathematical growth. It allows students who are beginning to form a mathematical structure to clarify it and those who have mastered isolated concepts to begin to see how these concepts might fit together.

At times you may find yourself tempted to introduce an activity by giving examples or modeling how the activity might be accomplished. Avoid this. If you do this, you rob students of the chance to approach the activity in their own individual way. Instead of making assumptions about what students will or won't do, watch and listen. The excitement and challenge of the activity—as well as the chance to work cooperatively—may bring out abilities in students that will surprise you.

USING THE MANIPULATIVES

Each activity in this book was written with a specific manipulative in mind. The six manipulatives used are: Geoboards, Color Tiles, Snap Cubes, Cuisenaire Rods, Pattern Blocks, and Tangrams. All of these are pictured on the cover of this book. If you are missing a specific manipulative, you may still be able to use the activity by substituting a different manipulative. For example, most Snap Cube activities can be performed with other connecting cubes. In fact, if the activity involves using the cubes as counters, you may be able to substitute a whole variety of manipulatives.

The use of manipulatives provides a perfect opportunity for authentic assessment. Watching how students work with the individual pieces can give you a sense of how they approach a mathematical problem. Their thinking can be "seen" in the way they use and arrange the manipulatives. When a class breaks up into small working groups, you can circulate, listen, and raise questions, all the while focusing on how your students are thinking.

Work with manipulatives often elicits unexpected abilities from students whose performance in more symbolic, number-oriented tasks may be weak. On the other hand, some students with good memories for numerical relationships have difficulty with spatial reasoning and can more readily learn from free exploration with hands-on materials. For all students, manipulatives provide concrete ways to tackle mathematical challenges and bring meaning to abstract ideas.

Overview of the Lessons

INVESTIGATING GROWTH PATTERNS

Backyard Improvements 9
Students use Pattern Blocks to build increasingly larger squares and staircase models, and examine the underlying number sequence patterns.

Ripples 14
Students investigate the pattern of growth that occurs when one Pattern Block is surrounded by blocks of the same kind, and each successive design is then surrounded.

Marquetry 19
Students use Snap Cubes to build models of Greek Cross numbers. They then build square arrays around their models, collect data regarding their designs, and look for patterns in the data.

The Pyramid Mystery 24
Using Snap Cubes, students build 3-dimensional models of pyramid numbers. They record data, look for patterns, and make predictions. They then build and investigate a sequence of structures in which each structure models a pair of pyramids joined base to base.

Bees in the Trees 29
Students use Snap Cubes to model ancestral trees of male bees and of their own families, leading to investigations of the Fibonacci sequence and an exponential growth sequence.

INVESTIGATING PATTERNS FROM GEOMETRY

Napkins and Place Mats 35
Students use Color Tiles to model a series of squares and rectangles. They then investigate the relationships that exist among the dimensions, the number of border tiles, and the number of interior tiles in each quadrilateral.

Pythagoras Delivers the Mail 40
Students examine squares built on the sides of right, obtuse, and acute Geoboard triangles. They look for relationships between the areas of the squares and the type of triangle on which they are built.

Polygons, Pegs, and Patterns 44
Students create a variety of polygons on their Geoboards, each having specified numbers of boundary pegs and interior pegs. They then find the areas of their polygons and search for patterns in their data.

The Airline Connection 48
Using circular Geoboards, students create polygons, make their diagonals, and look for a way to relate the number of diagonals that can be made in a polygon to the number of sides in the polygon. They then apply their discoveries to an application involving airline routing.

Patterns and Functions, Grades 7-8

Inside Out, Outside In 52
Using Pattern Blocks, students investigate the sums of the measures of the interior angles and of the exterior angles of a variety of polygons.

INVESTIGATING PATTERNS AND FUNCTIONS AS PROBLEM-SOLVING TOOLS

Count Square and Countess Triangle 58
Using Color Tiles, students search to find as many squares as they can on an 8-by-8 checkerboard. They then use green Pattern Block triangles to explore a similar problem involving equilateral triangles.

Greek Border Designs 63
Students determine how to calculate the number of Color Tiles needed for designs in a sequence without actually building each design. They then create their own sequence of designs based on given patterns.

Table for 63, Please 68
Students use Pattern Blocks to investigate how perimeter changes as blocks are added to a shape.

Birthday Cakes 73
Using a circular Geoboard, students look for patterns in the number of regions formed when points on the circumference of a circle are connected. They then extend their investigation to Geoboard rectangles.

INVESTIGATING OTHER TOPICS USING PATTERNS AND FUNCTIONS

Visual Effects 78
Students use Cuisenaire Rods and/or Tangram pieces to build successive terms of given sequences. They then create an original sequence, write rules describing it, exchange rules with another group, and try to build each other's sequences.

Pascal Pastimes 83
Students search for all possible paths that can be made from a corner peg on a Geoboard to each of the other pegs. They then perform a probability experiment using Color Tiles and investigate how the outcome is related to the results of the first activity.

Beehive Buzz 88
Students search to find combinations of Pattern Blocks that can be used to cover a honeycomb design made from 7 hexagons.

Carol's Kite Kits 93
Students use Snap Cubes to build increasingly larger models of kite frames. They gather data about the amount of materials needed to build each kite and look for underlying number patterns.

Investigating Growth Patterns

1. Backyard Improvements, page 9   (Pattern Blocks)
2. Ripples, page 14   (Pattern Blocks)
3. Marquetry, page 19   (Snap™ Cubes)
4. The Pyramid Mystery, page 24   (Snap™ Cubes)
5. Bees in the Trees, page 29   (Snap™ Cubes)

The lessons in this cluster present students with a variety of scenarios in which they can explore different types of patterns. Although no prior knowledge is assumed for any of the lessons, the sequences students encounter in *Backyard Improvements* may help them identify and describe some of the sequences generated in the other activities.

1. Backyard Improvements (Exploring number sequences)

In this activity, students are introduced to arithmetic sequences and the sequences formed by perfect squares and triangular numbers. No prior knowledge of these concepts is assumed, making the lesson ideal as an introductory activity.

In both *On Their Owns*, students experiment with writing algebraic expressions to generalize the patterns they find. Through test-and-check techniques, they can discover the formula for the sum of the first *n* counting numbers.

This lesson is written as a Pattern Blocks activity but could be done with Color Tiles and Cuisenaire Rods instead.

2. Ripples (Exploring patterns of growth)

This lesson provides students with opportunities to formulate and test predictions based on the growth of design sequences they build using Pattern Blocks. As the patterns of growth emerge, students are asked to represent the patterns algebraically.

Although no prior experience with growth sequences is called for in this activity, it may be easier for students to generalize their findings if they have had some experience with the types of sequences explored in *Backyard Improvements* (page 9).

3. Marquetry (Extending the exploration of number sequences)

In this activity, students discover the sequence of Greek Cross numbers by building and examining models of Greek Cross structures. They also explore the underlying number patterns generated by the different components of the structures.

Students are asked to generalize the patterns they find, and some knowledge of square and triangular numbers will prove helpful here. *Where's the Mathematics?* (page 21) provides complete explanations as to the derivations of the algebraic expressions representing the data for the *n*th Greek Cross design. If the writing of these expressions is beyond the students' ability, teachers may choose to ask them to, instead, describe their patterns verbally.

4. The Pyramid Mystery (Investigating relationships among sequences)

In this activity, students discover the sequence of pyramid numbers by building and examining models of pyramid (and double-pyramid) structures. They also explore the underlying number patterns generated by the different components of the structures.

As the number patterns emerge, students are asked to compare the different sequences to see how they are related. They also look to see if they recognize any familiar sequences among the data (square numbers, Greek Cross numbers, etc.), and examine how these sequences are related to the sequence of pyramid numbers.

5. Bees in the Trees (Exploring the Fibonacci and exponential growth sequences)

Through their investigation of ancestral trees, students have the opportunity to discover and learn about the Fibonacci and exponential growth sequences. In *For Their Portfolio*, students research and share examples of the Fibonacci sequence that occur in nature.

BACKYARD IMPROVEMENTS

- Pattern recognition
- Arithmetic sequences
- Square and triangular numbers
- Organizing and interpreting data

Getting Ready

What You'll Need

Pattern Blocks, 1 set per pair (only squares are needed)

Activity master, page 99

Overview

Students use Pattern Blocks to build increasingly larger squares and staircase models, and examine the underlying number sequence patterns. In this activity, students have the opportunity to:

- recognize patterns
- discover properties of arithmetic sequences
- learn about square and triangular numbers
- make predictions based on patterns
- compare sequences to see how they are related

Other *Super Source* activities that explore these and related concepts are:

Ripples, page 14

Marquetry, page 19

The Pyramid Mystery, page 24

Bees in the Trees, page 29

The Activity

On Their Own (Part 1)

Alan is planning to build a square patio in his backyard using square pieces of slate that measure 1 foot by 1 foot. He needs to decide what dimensions to make the patio so that he can determine how many pieces of slate to order. What information can you gather that might help Alan with his project?

- Use the orange Pattern Blocks to represent the square pieces of slate. Working with your partner, build models of increasingly larger square patios, beginning with the smallest possible patio, the one made from one square.

- Each time you build a new patio model, record the number of blocks you added to the previous model to build the next bigger square, the perimeter of the patio, and the total number of blocks in the new patio. Organize your data in a table.

- Record the data for the first ten squares, but build squares only until you discover a pattern that will produce all the numbers needed for your table.

- Look for relationships between the dimensions of the square patio, its perimeter, and the total number of pieces of slate needed. Think about how you might generalize your findings. For example, think about what the data would be for a patio with dimensions *n* feet by *n* feet.

- Be ready to discuss and justify your conclusions.

Thinking and Sharing

Have students help you create a class chart for the patios they modeled. Label the column headings: *side length of square patio, number of blocks added, perimeter,* and *total number of blocks*. If students have differing data, have them work together to come to agreement.

Use prompts like these to promote class discussion:

- What patterns do you see in the data?
- How would you describe the number sequence in the *number of blocks added* column?
- How would you describe the number sequence in the *perimeter* column?
- How would you describe the number sequence in the *total number of blocks* column?
- How many blocks would be needed for the 11th square? the 12th square? the *n*th square?
- How many blocks would be needed for the 20th square? How do you know?
- What algebraic expressions can you write to express the relationships you found?

On Their Own (Part 2)

What if... Alan also decides to build a set of stairs from one level of his backyard to another? He wants to use railroad ties for the steps, and must determine how many he needs. What information can you gather that might help Alan with this project?

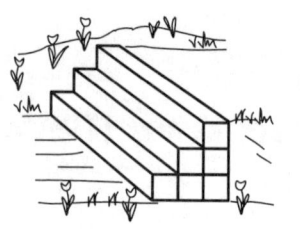

- Using the orange squares to represent the side view of the railroad ties, work with your partner to build models of increasingly larger staircases.

- Each time you build a new staircase, record the number of blocks you added to build the next taller step, the perimeter of the new side view, and the total number of ties needed to build the staircase. Organize your data in a table.

- Record the data for the first ten staircases, but build sets of stairs only until you discover a pattern that will produce all the numbers needed for your table.

- Look for relationships between the number of steps in the staircase, the perimeter of the side view, and the total number of railroad ties needed. Think about how you might generalize your findings to a staircase with *n* steps.

- Compare the sequences you found in the first activity with the sequences you found in this activity. Discuss your observations with your partner.

Thinking and Sharing

Have students help you create a chart similar to that in the first activity. Label the column headings: *number of steps in staircase, number of railroad ties added, perimeter,* and *total number of railroad ties*. Invite students to share their observations with the class.

Use prompts like these to promote class discussion:

- What patterns do you see in the data?
- How would you describe the number sequence generated by the entries in the *number of railroad ties added* column? the *perimeter* column? the *total number of railroad ties* column?
- How many railroad ties would be needed for the 11th staircase? the 12th staircase? the *n*th staircase?
- How many railroad ties would be needed for the 20th staircase? How do you know?
- What algebraic expressions can you write to express the the relationships you found?
- What did you notice when you compared the sequences from Part 1 with those from Part 2?

For Their Portfolio

Describe how the various sequences you found in these activities were alike and how they were different. Make up some sequences of your own that are similar in some way to the ones that were generated by the data.

Teacher Talk

Where's the Mathematics?

The data students collect provide them with an opportunity to discover several types of sequences. The entries in the first column of each table form the sequence of consecutive positive integers, an example of an arithmetic sequence. Help students to see that each new term of an arithmetic sequence is generated by adding the same value to the previous term. In this case, that value (also called the *common difference*) is 1. The entries in the *perimeter* column of both tables also form an arithmetic sequence where the common difference between any two consecutive terms is 4. Students should notice that each term in this column is four times the number in the first column; therefore, the *n*th term in this column could be represented using the algebraic expression 4*n*.

The table here shows the data pertaining to the materials needed to build square patios.

side length of square patio	number of blocks added	perimeter	total number of blocks
1	1	4	1
2	3	8	4
3	5	12	9
4	7	16	16
5	9	20	25
6	11	24	36
7	13	28	49
8	15	32	64
9	17	36	81
10	19	40	100
⋮	⋮	⋮	⋮
n	2*n* – 1	4*n*	*n* x *n*, or *n*²

©Cuisenaire Company of America, Inc.   BACKYARD IMPROVEMENTS ◆ Patterns/Functions ◆ Grades 7-8

The entries in the second column (1, 3, 5, 7, 9, ...) also form an arithmetic sequence. The common difference between any two consecutive terms is 2. Students may recognize this sequence as the sequence of positive odd integers. Each number in this sequence is 1 less than twice the side length of the corresponding square. Thus, to find the value of the nth term in this sequence, the algebraic expression $2n - 1$ can be used.

The sequence 1, 4, 9, 16, 25, 36, ... , generated by the data in the last column, is the sequence of perfect squares. Students should recognize that since the length and width of a square are always equal, and the total number of pieces of slate needed to build a square patio is the product of the length and width, that total will necessarily be a square number. Thus, to find the number of pieces of slate needed to build a patio with side length of n feet, calculate n^2.

It is important to point out that the sequence of perfect squares is not an arithmetic sequence; the difference between terms is not constant. However, the sequence formed by these differences *is* arithmetic, an interesting and intriguing characteristic of the sequence of perfect squares. Students will have opportunities to learn more about sequences that behave this way when they study them later on in high school algebra.

As students look for relationships among the sequences in the table and consider the relationship between the values in the 2nd and 4th columns, they may discover that square numbers can always be generated by adding consecutive odd numbers beginning with 1. For example, adding the first four odd integers (1 + 3 + 5 + 7) gives 16, or 4^2; adding the first seven odd integers (1 + 3 + 5 + 7 + 9 + 11 + 13) gives 49, or 7^2. This interesting fact may prompt students to explore other patterns related to perfect squares.

The table below shows the data pertaining to the materials needed to build staircases.

number of steps in staircase	number of railroad ties added	perimeter	total number of railroad ties
1	1	4	1
2	2	8	3
3	3	12	6
4	4	16	10
5	5	20	15
6	6	24	21
7	7	28	28
8	8	32	36
9	9	36	45
10	10	40	55
⋮	⋮	⋮	⋮
n	n	4n	$\frac{n(n+1)}{2}$

Students should discover that with each new step, the number of railroad ties added is 1 larger than the previous number added. They may also notice that, for example, if it takes 5 more railroad ties to build the 5-step staircase than to build the 4-step staircase, then it will take 6 more ties to build the 6-step staircase from the 5-step staircase. Algebraically speaking, to build an n-step staircase, n additional railroad ties will need to be added to the previous structure.

To find the total number of railroad ties needed to build a set of stairs, students may resort to counting the ties each time they add an entry to the table. Some students may realize that they are simply adding the consecutive integers from 1 up to and including the number of ties they've added to form the new staircase. For example, to build a 7-step staircase, the total number of railroad ties needed will be 1 + 2 + 3 + 4 + 5 + 6 + 7, or 28. Finding a way to represent such a sum algebraically may prove challenging for some students. Some students may suggest the expression $n + (n - 1) + (n - 2) + (n - 3) + ... + 1$. Others, through trial-and-error techniques, may be able to discover the relationship between n and the sum of the first n counting numbers and derive the expression for the sum: $\frac{n \times (n + 1)}{2}$. Some students may need to be given this formula and asked to confirm it using their their data.

The sequence generated by the entries in the *total number of railroad ties* column (1, 3, 6, 10, 15, 21, ...) is known as the sequence of triangular numbers. Each term is the sum of consecutive integers. Beyond that, it is interesting to note that by adding together any two consecutive terms of the triangular sequence, a term from the sequence of perfect squares is generated. The triangular numbers can also be generated by using counters (or other manipulatives) to build triangular arrays as shown below; hence the name "triangular" numbers.

triangular numbers as
equilateral triangle arrays

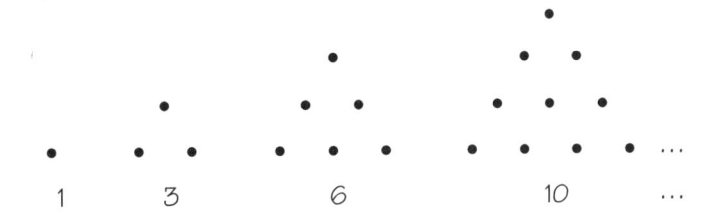

1   3   6   10   ...

triangular numbers as
right triangle arrays

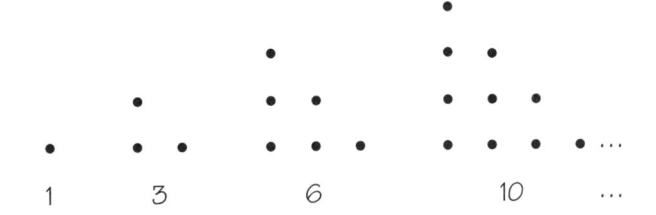

1   3   6   10   ...

©Cuisenaire Company of America, Inc.   BACKYARD IMPROVEMENTS ◆ Patterns/Functions ◆ Grades 7-8   **13**

RIPPLES

- Pattern recognition
- Growth patterns
- Arithmetic sequences
- Writing algebraic expressions

Getting Ready

What You'll Need

Pattern Blocks, 1 set per pair

Colored pencils or markers

Activity master, page 100

Overview

Students investigate the pattern of growth that occurs when one Pattern Block is surrounded by blocks of the same kind, and each successive design is then surrounded. In this activity, students have the opportunity to:

- discover patterns
- make predictions
- investigate arithmetic sequences
- use algebraic expressions to generalize patterns

Other *Super Source* activities that explore these and related concepts are:

Backyard Improvements, page 9

Marquetry, page 19

The Pyramid Mystery, page 24

Bees in the Trees, page 29

The Activity

On Their Own (Part 1)

Mariano notices that when a stone is thrown into a calm body of water, it produces a ripple effect of larger and larger circles whose centers are the stone's point of contact with the water's surface. Mariano wants to investigate a similar kind of ripple effect using Pattern Block shapes. What patterns can be discovered in a sequence of ripples?

- Work with a partner. Using the Pattern Block square or the blue or tan rhombus as the "stone," completely surround your stone with blocks of the same shape to form the first "ripple." Record your design. Color it using one color for the original stone and another color for the blocks in the first ripple.

- Surround your first ripple design with more blocks of the same shape to form the second ripple. Make sure that the entire perimeter of the ripple is surrounded with new blocks. Record your new design. Color the new ripple with a different color.

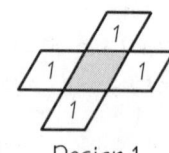

Design 1

Design 2

- Predict how many stones it will take to form the third ripple and check your prediction.
- As each new ripple is generated, record the number of blocks in the new ripple, the total number of blocks in the ripple design, and the perimeter of the ripple design.
- Continue predicting, surrounding your designs, and recording the data until you have built the sixth ripple.
- Look for patterns in your results. See if you can write algebraic expressions that generalize your findings.
- Repeat the activity using yellow Pattern Block hexagons.

Thinking and Sharing

Ask students who used the square for their stone to work together to prepare a class chart showing their data. Do the same for pairs who used the blue rhombus and those who used the tan rhombus. Also ask some pairs to work on a class chart showing the data for the hexagon. Have all four charts available for the class discussion.

Use prompts like these to promote class discussion:

- What do you notice about the data in the class charts?
- What did you notice about the numbers of blocks needed to surround the different-shaped stones and their ripples?
- How would you describe the patterns generated by the entries in the different columns of the charts?
- What relationship do you see between the number of sides on the original stone shape and the growth pattern of its ripples?
- How did you go about making your predictions?
- What predictions would you make for the 10th ripple for each shape?
- What algebraic expressions did you write to express the relationships you found?

On Their Own (Part 2)

What if... Mariano decides to investigate the ripple patterns for the triangle and trapezoid stones? Will he find that these shapes produce patterns similar to the ones he found for the other shapes?

- Using the procedure from the first part of the activity, gather data about ripples formed from using the Pattern Block triangle as the stone.
- Compare the patterns formed by the triangle data with those you found in Part 1.
- Predict what will happen if you use the Pattern Block trapezoid for the stone. Test your prediction by building the first three ripple designs. As you add trapezoids to the design, be sure to join short edge to short edge and long edge to long edge.
- Discuss whether or not the trapezoid designs follow growth patterns that are similar to those of the other shapes. Be ready to justify your findings.

Thinking and Sharing

Ask students to add the charts for the triangle and trapezoid shapes to those already displayed. Students may have difficulty in completing the chart for the trapezoid, but ask them to fill in as much as possible.

Use prompts like these to promote class discussion:

- What patterns did you find in your data?
- What similarities/differences exist between the data from the first activity and that from the second? How can you explain this?
- What happened when you tried to build ripple designs around the trapezoid stone? Why do you think this happened?
- What predictions would you make for the 10th ripple in the triangle sequence? the 10th ripple in the trapezoid sequence?

For Their Portfolio

Write a letter to Mariano explaining how the shape of a stone affects the patterns formed by its ripples. Use examples and drawings to help illustrate your explanation.

Teacher Talk

Where's the Mathematics?

Students enjoy having the opportunity to formulate and test predictions based on their findings. The patterns in the sequences of numbers generated by the ripples emerge almost immediately and enable students to predict, check, and verify their findings for the square (or rhombus) and hexagon. The charts for these shapes are shown below.

square or rhombus

ripples	number of blocks added	total number of blocks	perimeter
1	4	5	12
2	8	13	20
3	12	25	28
4	16	41	36
5	20	61	44
6	24	85	52
⋮	⋮	⋮	⋮
n	$4n$	$4\left[\dfrac{n(n+1)}{2}\right]+1$	$8n+4$

hexagon

ripples	number of blocks added	total number of blocks	perimeter
1	6	7	18
2	12	19	30
3	18	37	42
4	24	61	54
5	30	91	66
6	36	127	78
⋮	⋮	⋮	⋮
n	$6n$	$6\left[\dfrac{n(n+1)}{2}\right]+1$	$12n+6$

Students should recognize that the number of sides on each shape determines how many blocks will be required to surround it the first time. The second ripple requires twice as many; the third ripple, three times as many, and so on.

As students experiment with algebraic expressions that will generate specific terms in a particular sequence, they may recognize that many of the patterns form arithmetic sequences. For the square (or rhombus), the entries in the *number of blocks added* column forms an arithmetic sequence in which the common difference is 4, while the entries in the same column for the hexagon form an arithmetic sequence whose common difference is 6. Students may also notice that these entries are multiples of the number of the ripple under consideration. They may see that to find the number of blocks added for a ripple around the square or rhombus, they can multiply the ripple number by 4. To find the number of blocks added for a ripple around a hexagon, they can multiply the ripple number by 6.

The entries in the *perimeter* column for the square (or rhombus) and the hexagon also form arithmetic sequences whose common differences are 8 and 12, respectively. Encourage students to search for relationships between the ripple number and the perimeter. If, for example, they identify the common difference in the sequence of perimeters for the square to be 8, they may, through test-and-check techniques, be able to determine that the perimeter in each case is 4 more than 8 times the number of ripples. Thus, the perimeter of a design with n ripples will be $8n + 4$. Similarly, the perimeter of designs built around the hexagon can be expressed as $12n + 6$, where n is, again, the number of ripples. Note that both of these expressions are equivalent to the general expression $s(2n + 1)$, where s is the number of sides of the original polygon.

Finding an algebraic expression that generates the entries in the *total number of blocks* columns may prove challenging for many students. Some students may notice that the terms in these sequences are each 1 more than a multiple of 4 (in the case of the quadrilaterals) and 1 more than a multiple of 6 (in the case of the hexagon), the 1 accounting for the original stone. This may prompt them to investigate the multipliers of 4 and 6 being used. Interestingly, these multipliers are the triangular numbers: 1, 3, 6, 10, ... (See *Backyard Improvements*, page 9, for more on triangular numbers.) Each triangular number is the sum of consecutive counting numbers. Students may be familiar with the formula for the sum of the first n counting numbers, $\frac{n \times (n + 1)}{2}$, and may be able to incorporate it into an expression representing the total number of blocks needed to build a design with n ripples surrounding, for example, a hexagon: $6 \times \left[\frac{n \times (n + 1)}{2}\right] + 1$. Some students may need help deriving this expression, but can be asked to use their data to verify it.

Some different occurrences emerge in Part 2 of the activity. The chart for the triangle is shown below.

triangle

ripples	number of blocks added	total number of blocks	perimeter
1	3	4	6
2	6	10	12
3	9	19	15
4	12	31	21
5	15	46	24
6	18	64	30
⋮	⋮	⋮	⋮
n	$3n$	$3\left[\frac{n(n+1)}{2}\right] + 1$	—

Many of the same patterns and types of sequences found in Part 1 exist in this chart, with the exception of the entries in the *perimeter* column. Students may notice that if the ripple number is even, 6 was added to the previous entry, while if the ripple number is odd, the value added was 3. Students may be able to discover why these values increase in an alternating pattern. For example, when the single triangle is surrounded, each of the surrounding blocks covers only one side of the original block, leaving 2 exposed sides on the new outer perimeter. It will take 6 triangles to surround these exposed sides for the second ripple. However, to surround the second ripple, there are three places where a single triangle will cover exposed sides of two different triangles. Therefore, 3 fewer triangles will be needed. This construction pattern continues for every pair of consecutive ripples.

When students attempt to build ripples around the trapezoid, they may experience confusion about knowing how to place the blocks. Even as early as in the first ripple, students will discover that there are several ways of placing the surrounding blocks.

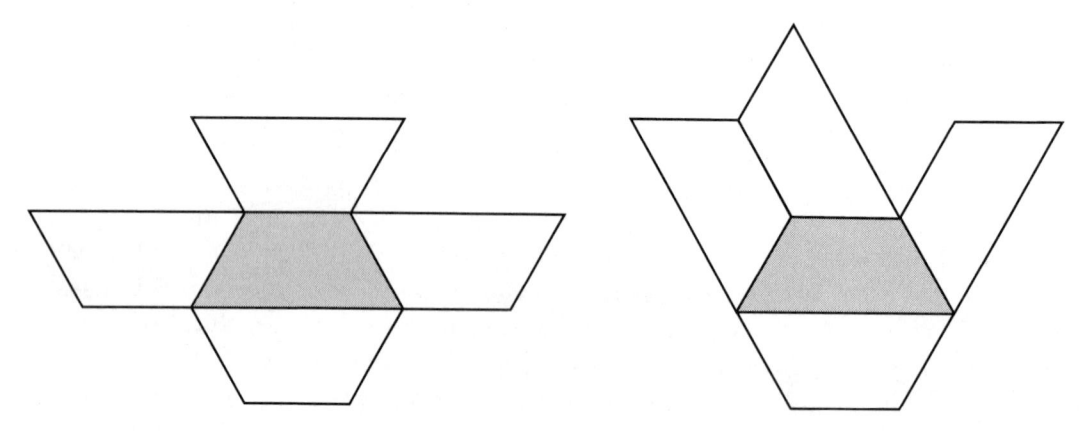

Two possible arrangements for first ripple

There are even more options for arrangements for the second ripple, and students may discover that not all second ripples require the same number of blocks. Given this situation, no consistent number patterns will emerge in the trapezoid data.

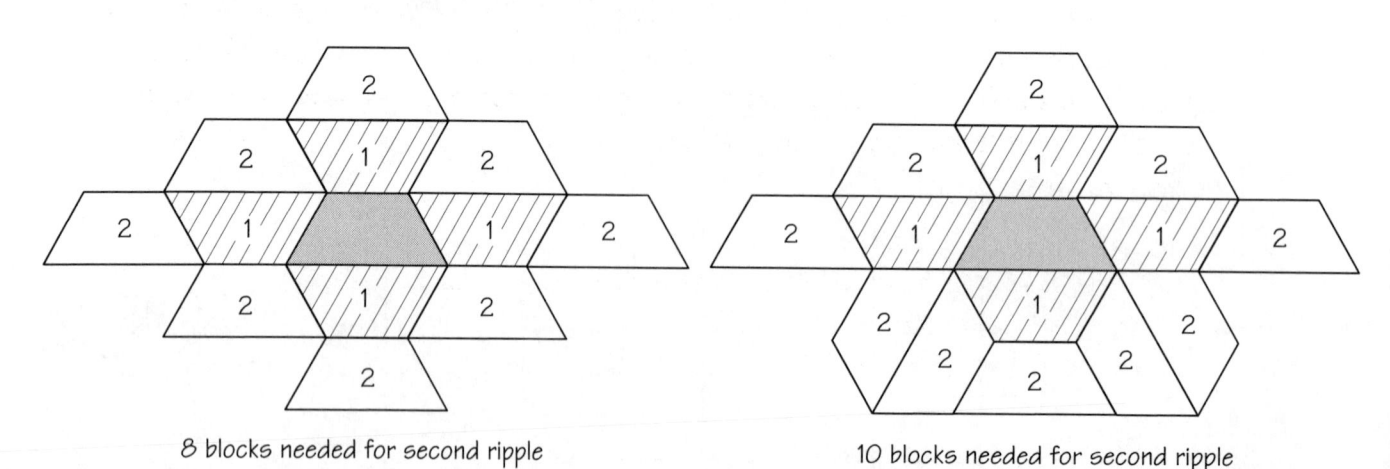

8 blocks needed for second ripple

10 blocks needed for second ripple

MARQUETRY

- Growth patterns
- Greek Cross numbers
- Pattern recognition
- Writing algebraic expressions

Getting Ready

What You'll Need

Snap Cubes, about 100 each of two colors per pair

Calculators (optional)

Activity master, page 101

Overview

Students use Snap Cubes to build models of Greek Cross numbers. They then build square arrays around their models, collect data regarding their designs, and look for patterns in the data. In this activity, students have the opportunity to:

- represent a number sequence geometrically
- collect, organize, and analyze data
- use patterns to make predictions
- investigate relationships among different sequences
- express relationships algebraically

Other *Super Source* activities that explore these and related concepts are:

Backyard Improvements, page 9
Ripples, page 14
The Pyramid Mystery, page 24
Bees in the Trees, page 29

The Activity

On Their Own (Part 1)

Marquetry is a technique in which pieces of different types of wood are inlaid into the surface of a piece of furniture to create a design. One popular design often used to decorate table tops is the Greek Cross design made from cubes of wood of varying colors. This type of design is not only interesting to look at, but is also full of hidden patterns. Can you find them?

- Using Snap Cubes to represent the cubes of wood, work with a partner to build the Greek Cross designs shown. Use only two colors (white and red, for example) for your designs, alternating colors as you build each larger design.

- For each Greek Cross design, record the design number, the number of cubes added to the previous structure, the number of cubes of each color, and the total number of cubes in the structure. Organize your data in a chart.

- Predict the number of cubes of each color that would be needed to build the fourth Greek Cross structure and check your prediction by building it. Do the same for the fifth structure.

- Look for patterns in your data. Use the patterns to continue your chart through the 10th Greek Cross design. Then generalize your findings by writing algebraic expressions for the data for the *n*th Greek Cross design, where possible.

Thinking and Sharing

Have students help you create a class chart displaying the data. Possible column headings might include: *design number, number of cubes added, number of (white) cubes, number of (red) cubes,* and *total number of cubes* in *Greek Cross*.

Use prompts like these to promote class discussion:

- What did you notice as you built successively larger Greek Cross structures?
- What patterns did you notice in the data?
- How accurate were your predictions?
- How did you generate the data for the the larger Greek Cross designs?
- What algebraic expressions did you write to generalize the patterns you found?

Explain to students that the numbers of cubes contained in their structures are called *Greek Cross numbers*. They are so named because the structures look like crosses and Greek mathematicians were the first to study the sequence of numbers in depth.

On Their Own (Part 2)

What if... the corner areas of the Greek Cross designs are filled in with wooden cubes? How would this affect the numbers of the different-colored cubes needed for each design?

- Using your models from the first activity, fill in the open "corners" of each Greek Cross with Snap Cubes to create a square. In each case, use the color that would come next in the alternating sequence.

- For each marquetry design, determine the dimensions of the square, the number of cubes added at each corner, the total number (and color) of cubes added, and the total number of cubes in the square design. Add columns to the chart you made in the first activity to record your data.

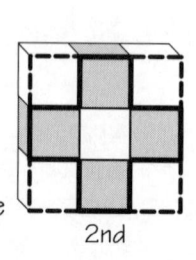

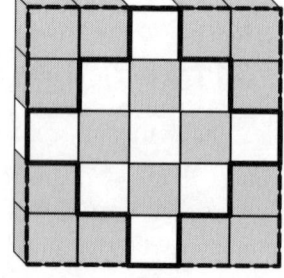

- Look for patterns in your data. Use the patterns to continue your chart through the 10th design. Then generalize your findings by writing algebraic expressions for the data for the *n*th design.

- Compare the different sequences formed by the entries in you chart. Discuss them with your partner and be ready to talk about any relationships you find.

Thinking and Sharing

Invite students to add their new data to the class chart. Have them label the new column headings as they proceed.

Use prompts like these to promote class discussion:
- What did you discover about the dimensions of the squares that you built surrounding the Greek Crosses? Why do you think there are no squares with even dimensions?
- What patterns did you notice in the data?
- What algebraic expressions did you write to generalize the patterns you found?
- Did you recognize any familiar sequences of numbers in any of the columns? If so, what were they?
- What other relationships did you discover among the sequences formed by the data?

For Their Portfolio

Picture a 25-inch by 25-inch square marquetry design made from cubes of wood measuring 1 inch on each edge. If the design is to contain the largest Greek Cross that can be fitted in the square, what materials will be needed? Explain your reasoning and show your calculations.

Teacher Talk

Where's the Mathematics?

There are numerous relationships that students may come across as they gather data from their Greek Cross structures. The chart below shows the data from the first activity, with the last column containing the numbers referred to as *Greek Cross numbers*.

design number	number of cubes added	number of (white) cubes	number of (red) cubes	total number of cubes in Greek Cross
1		1	0	1
2	4	1	4	5
3	8	9	4	13
4	12	9	16	25
5	16	25	16	41
6	20	25	36	61
7	24	49	36	85
8	28	49	64	113
9	32	81	64	145
10	36	81	100	181
⋮	⋮	⋮	⋮	⋮
n	$4(n-1)$	——	——	$n^2 + (n-1)^2$

©Cuisenaire Company of America, Inc.   MARQUETRY ◆ Patterns/Functions ◆ Grades 7-8

The second column, *number of cubes added*, contains entries that are consecutive multiples of 4. After building three or four Greek Cross structures, students may begin to notice that each new structure is formed by adding cubes in increasing multiples of 4. For example, the second Greek Cross is made up of 1 + 4 cubes, the third Greek Cross is made up of 1 + 4 + 8 cubes, and the fourth, of 1 + 4 + 8 + 12 cubes. Students may suggest that by subtracting 1 from the design number and multiplying this result by 4, they can determine the number of cubes that were added. For example, for the fifth structure, (5 − 1) x 4, or 16, cubes were added. In general, for the *n*th design, 4 x (*n* − 1) cubes will need to be added.

Students may recognize the entries that appear in the third and fourth columns as numbers from the sequence of perfect squares. The total number of white (first color) cubes in each Greek Cross structure can be represented by the square of an odd integer, $1^2, 3^2, 5^2, 7^2, ...$, while the total number of red (second color) cubes can be represented by the square of an even integer, $2^2, 4^2, 6^2, 8^2, ...$. In fact, the total number of cubes needed in each design, the Greek Cross numbers, is always the sum of two consecutive terms from the sequence of perfect squares. In general, the total number of cubes in the *n*th Greek Cross structure is $n^2 + (n − 1)^2$.

The chart for the second activity is shown below and should be used in conjunction with the first chart.

design number	dimensions of square array	number of cubes added at each corner	total number (and color) of cubes added to the corners	total number of cubes in square
1	1 x 1	0	0	1
2	3 x 3	1	4 (white)	9
3	5 x 5	3	12 (red)	25
4	7 x 7	6	24 (white)	49
5	9 x 9	10	40 (red)	81
6	11 x 11	15	60 (white)	121
7	13 x 13	21	84 (red)	169
8	15 x 15	28	112 (white)	225
9	17 x 17	36	144 (red)	289
10	19 x 19	45	180 (white)	361
⋮	⋮	⋮	⋮	⋮
n	(2*n* − 1) x (2*n* − 1)	$\frac{n(n-1)}{2}$	$4\left[\frac{n(n-1)}{2}\right]$ or $2n(n-1)$	$(2n-1)^2$

As students start enclosing the Greek Cross structures in squares, they will notice that the dimensions of the square depend upon the longest row in the Greek Cross. This row always contains an odd number of cubes, 1 less than twice the design number. Thus, the dimensions can be represented by the expression (2*n* − 1) x (2*n* − 1), where *n* is the design number.

Beginning with the second design, the entries in the *number of cubes added at each corner* column form the triangular number sequence: 1, 3, 6, 10, 15, 21, ... (See *Backyard Improvements*, page 9, for more on triangular numbers.) This sequence can be generated using the expression $\frac{n \times (n + 1)}{2}$ for consecutive integer values of *n* beginning with 1; but to compensate for the fact that the sequence begins at the second design, $n - 1$ must be substituted for *n* in the expression. Thus, the entries in the column can be represented by the expression $\frac{(n - 1) \times n}{2}$, where *n* is the design number. Students should recognize that to find the total number of corner cubes needed to complete the square, this value must be multiplied by 4.

Students may notice that the last column in the chart is the sequence of the squares of odd numbers. They may also notice, when comparing the *total number of cubes added to the corners* column to the *total number of cubes in the Greek Cross* column, that the corresponding entries in these two columns differ by 1 and that the sequences themselves "grow" in the same way. Furthermore, when the two corresponding numbers from these sequences are added together, the result is the perfect square representing the total number of cubes needed for the entire design.

THE PYRAMID MYSTERY

- Growth patterns
- Pyramid numbers
- Pattern recognition

Getting Ready

What You'll Need

Snap Cubes, about 60 each of two different colors per pair

Calculators, 1 per pair

Activity master, page 102

Overview

Using Snap Cubes, students build 3-dimensional models of pyramid numbers. They record data, look for patterns, and make predictions. They then build and investigate a sequence of structures in which each structure models a pair of pyramids joined base to base. In this activity, students have the opportunity to:

- represent a number sequence geometrically
- collect, organize, and analyze data
- use patterns to make predictions
- investigate relationships among different sequences

Other *Super Source* activities that explore these and related concepts are:

Backyard Improvements, page 9

Ripples, page 14

Marquetry, page 19

Bees in the Trees, page 29

The Activity

On Their Own (Part 1)

The Egyptians were known for their magnificent granite and limestone pyramids and the remarkably mysterious methods used in building them. Can you build models of the pyramids and unlock the mystery surrounding the number of blocks needed for their construction?

- Using Snap Cubes to represent blocks of granite and limestone, work with a partner to build pyramids like those shown below. Use only two colors (yellow and brown, for example) for your pyramids. Alternate colors, so that each new outer layer of cubes is the opposite color of the layer underneath.

- For each structure, record its number in the sequence of pyramids, the number of cubes added to the previous structure, the number of cubes of each color, and the total number of cubes in the pyramid. Organize your data in a chart.

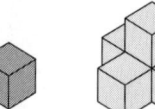

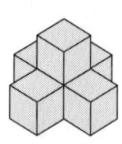

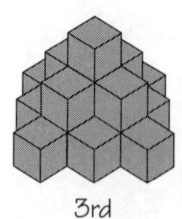

  1st   2nd   3rd

- Predict what cubes would be needed to build the fourth pyramid. Check your prediction by building the fourth pyramid.

24 the Super Source™ ◆ Patterns/Functions ◆ Grades 7-8   ©Cuisenaire Company of America, Inc.

- Look for patterns in your data. Use the patterns to continue your chart through the 10th pyramid.
- Compare the sequences formed by the entries in the different columns of your chart. Be ready to talk about your observations and any relationships you find.

Thinking and Sharing

Have students help you create a class chart displaying the data. Possible column headings might include: *number of pyramid, number of cubes added, number of (yellow) cubes, number of (brown) cubes,* and *total number of cubes in pyramid.*

Use prompts like these to promote class discussion:

- What did you notice as you built successively larger pyramids?
- How accurate were your predictions about the fourth pyramid?
- How did you generate the data for the larger pyramids?
- What patterns and/or sequences were you able to identify in the data?
- How would you calculate the total number of cubes needed to build the 20th pyramid?
- What relationships did you find among the number sequences generated by the data?

Explain to students that the numbers of cubes contained in their structures are called *pyramid numbers*.

On Their Own (Part 2)

What if... *the sequence of structures was generated by surrounding each previous structure on all sides with cubes of a contrasting color? How would this affect the numbers of different-colored cubes needed for each structure?*

- Using Snap Cubes, work with a partner to build "octamids" like those shown below. Use the same color arrangements as you used in the first activity.
- For each structure, record its number in the sequence of octamids, the number of cubes in its outer shell, the number of cubes of each color, and the total number of cubes in the octamid. Organize your data in a chart.
- Look for patterns in your data. Use the patterns to continue your chart through the 10th octamid.
- Using the charts you made in the two activities, look for patterns and relationships among the sequences formed by the data. Be ready to discuss your findings.

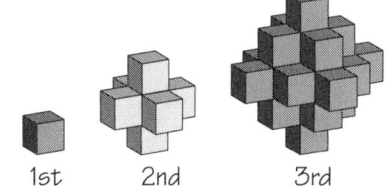

Thinking and Sharing

Create a class chart similar to the chart from the first activity and have students share their data and discuss the patterns they used to help complete their charts. Have them compare the data from the two activities and discuss their observations.

Use prompts like these to promote class discussion:

- What did you notice as you built successively larger octamids?
- How are octamids similar to pyramids?
- Did you use the pyramid structures to help build the octamids? If so, explain.
- How did you generate the data for the larger octamids?
- Did you use data from the first chart to help in completing the second chart? If so, explain.
- How would you calculate the total number of cubes needed to build the 20th octamid?
- What patterns and/or sequences were you able to identify in the data?
- What relationships did you find among the number sequences generated by the data?

For Their Portfolio

Write a short paragraph or two describing the relationship between the Greek Cross numbers and the Pyramid numbers. Explain the relationship both geometrically and numerically, using diagrams where helpful to illustrate your explanations.

Teacher Talk

Where's the Mathematics?

The data produced by the activities in this lesson provide students with an opportunity to see how several of the special "named" sequences that they've studied are related.

The chart below shows the data for the first ten Snap Cube pyramids.

number of pyramid	number of cubes added	number of (yellow) cubes	number of (brown) cubes	total number of cubes in pyramid
1	(1)	0	1	1
2	5	5	1	6
3	13	5	14	19
4	25	30	14	44
5	41	30	55	85
6	61	91	55	146
7	85	91	140	231
8	113	204	140	344
9	145	204	285	489
10	181	385	285	670

The second column, *number of cubes added*, contains the Greek Cross numbers: (1), 5, 13, 25, 41, 61, 85, ... (See *Marquetry*, page 19, for more on Greek Cross numbers.) Students may notice that each new term of this sequence increases by an increasing multiple of 4. For example:

$$5 = 1 + 4 \qquad 13 = 1 + 4 + 8 \qquad 25 = 1 + 4 + 8 + 12 \qquad 41 = 1 + 4 + 8 + 12 + 16$$

Other students may have generated the entries in this column by observing that each number is the sum of the squares of two consecutive integers. The second entry is $2^2 + 1^2$; the third entry is $3^2 + 2^2$, etc. In general, the number of cubes added for the *n*th pyramid is $n^2 + (n-1)^2$.

A third method for finding the number of cubes added is to look at the bottom layer of the pyramid structure. The bottom layer shows rows of consecutive odd numbers of cubes. For example, in the second pyramid, the bottom layer consists of rows containing either 1 or 3 (the first two odd numbers) cubes, adding 1 + 3 + 1, or 5, new cubes. The third pyramid consists of rows containing 1, 3, or 5 (the first three odd numbers) of cubes, adding 1 + 3 + 5 + 3 + 1, or 13, new cubes. Students can use this notion of a "symmetrical sum of odd integers" to generate the other entries in this column.

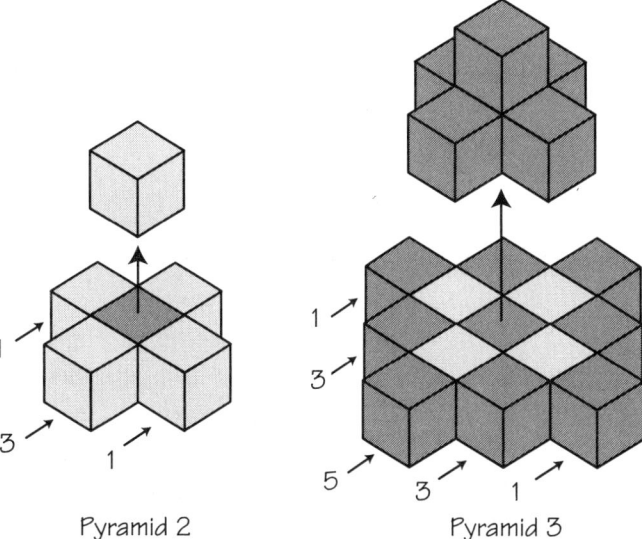

Pyramid 2   Pyramid 3

The third and fourth columns (listing the numbers of yellow and brown cubes in each pyramid structure), when looked at in combination, yield a sequence: 0, 1, 5, 14, 30, 55, 91, ... Although students may at first have trouble identifying a pattern here, if they calculate the difference between any two consecutive terms of this sequence, they will discover that the number is always a perfect square: 1 − 0 = 1, 5 − 1 = 4, 14 − 5 = 9, and so on. They can then continue adding consecutive perfect squares to find the number of yellow cubes and brown cubes used in each pyramid.

To find the total number of cubes in the pyramid, students may decide to simply add the numbers of yellow and brown cubes, or they may choose to combine the number of cubes added for that pyramid to the total used in the previous structure. A third method involves slicing the pyramid into "vertical staircases," as shown below, and adding the number of cubes used in each staircase. The numbers of cubes in the staircases are consecutive perfect squares. For example, the third pyramid consists of 1 + 4 + 9 + 4 + 1, or $1^2 + 2^2 + 3^2 + 2^2 + 1^2$, or 19, cubes. By continuing this pattern, students can calculate the number of cubes needed to build the 20th pyramid using the following symmetrical sum of consecutive squares: $1^2 + 2^2 + 3^2 + ... + 19^2 + 20^2 + 19^2 + ... + 3^2 + 2^2 + 1^2$, giving a total of 5340 cubes.

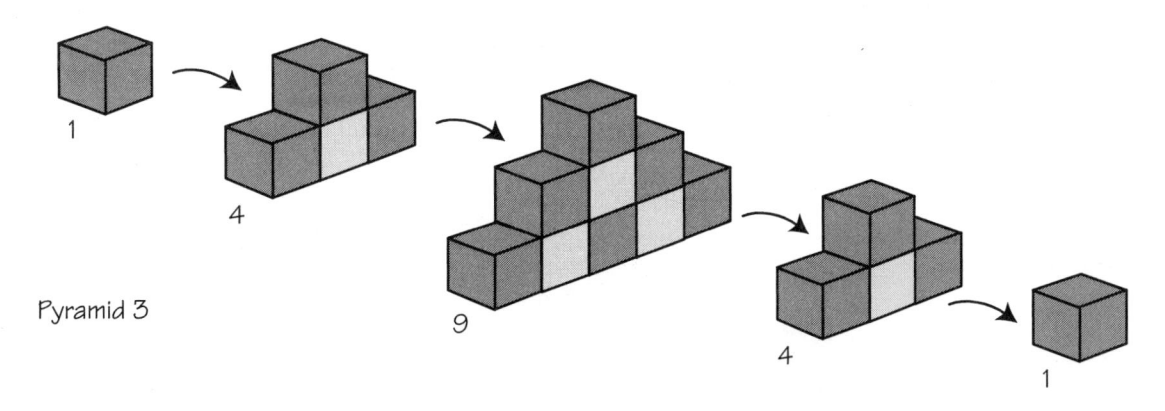

Pyramid 3

The data for the pyramid structure can be extremely useful in completing the chart for the "octamids."

number of octamid	number of cubes in new outer shell	number of (yellow) cubes	number of (brown) cubes	total number of cubes in octamid
1	(1)	0	1	1
2	6	6	1	7
3	18	6	19	25
4	38	44	19	63
5	66	44	85	129
6	102	146	85	231
7	146	146	231	377
8	198	344	231	575
9	258	344	489	833
10	326	670	489	1159

The second column, *number of cubes in new outer shell*, contains entries that are the result of adding two consecutive Greek Cross numbers. For example, 6 = 5 + 1, 18 = 13 + 5, 38 = 25 + 13, and 66 = 41 + 25. Thus, to find the number of cubes in the outer shell of the sixth octamid, students can add the Greek Cross number found in the second column of the first chart for the sixth pyramid (61) to the Greek Cross number preceding it (41) to get 102. In general, to find the number of cubes in the outer shell for the *n*th octamid, students can combine the number of cubes added for the *n*th pyramid and the (*n* – 1)th pyramid. This should seem reasonable, because the *n*th octamid is made up of the *n*th pyramid with the inverted (*n* – 1)th pyramid beneath it.

The third and fourth columns, listing the numbers of yellow and brown cubes in each octamid structure, when looked at in combination, yield the pyramid sequence: (0), 1, 6, 19, 44, 85, 146, ... When students calculate the difference between any two consecutive terms of this sequence, they may be surprised to discover that the number is always a Greek Cross number. For example, 1 – 0 = 1, 6 – 1 = 5, 19 – 6 = 13, and so on. The relationships among these sequences can help students find the numbers of yellow and brown cubes for larger octamids.

In order to find the total number of cubes in each octamid, students may decide to simply add the numbers of yellow and brown cubes, or they may choose to combine the number of cubes in the outer shell for each structure with the total number of cubes used in the previous octamid. A third method involves slicing the octamid into vertical Greek Crosses, as shown below, and adding the number of cubes in each Greek Cross. For example, the third octamid consists of 1 + 5 + 13 + 5 + 1, or 25, cubes; the fourth, of 1 + 5 + 13 + 25 + 13 + 5 + 1, or 63, cubes. Students can use these symmetrical sums of Greek Cross numbers to calculate the number of cubes needed to build an octamid of any size.

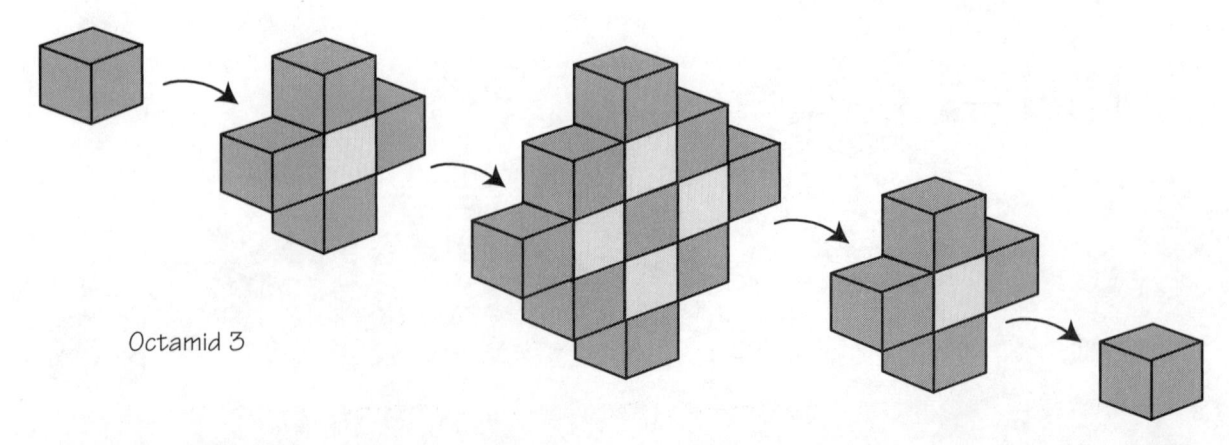

Octamid 3

BEES IN THE TREES

- Growth patterns
- Fibonacci sequence
- Exponential growth sequence
- Exponential notation

Getting Ready

What You'll Need

Snap Cubes, 35 each of black, yellow, blue, and purple, per pair

Toothpicks, approximately 100 per pair

Activity master, page 103

Overview

Students use Snap Cubes to model ancestral trees of male bees and of their own families, leading to investigations of the Fibonacci sequence and an exponential growth sequence. In this activity, students have the opportunity to:

- identify, extend, and compare growth patterns
- make predictions based on patterns
- learn about different types of sequences
- use exponential notation to rename expressions
- collect, organize, and analyze data

Other *Super Source* activities that explore these and related concepts are:

Backyard Improvements, page 9

Ripples, page 14

Marquetry, page 19

The Pyramid Mystery, page 24

The Activity

On Their Own (Part 1)

The family tree of a male bee is both unusual and interesting. The male bee is created through a process known as parthenogenesis, whereby he has a single parent, only a mother. The female bee, however, has both a mother and a father. What patterns can you find by tracing the ancestry of a male bee?

- Work with a partner. Construct a family tree of a male bee using black Snap Cubes to represent male bees, yellow Snap Cubes to represent female bees, and toothpicks to represent connections from generation to generation. Here's how:

  ◆ Place a black Snap Cube on your work surface to represent the male bee. Place a toothpick "connecting" this cube to a yellow Snap Cube representing the male bee's mother.

  ◆ From the mother bee, use 2 toothpicks – one leading to a black cube, the other to a yellow cube – to show the connection to her two parents.

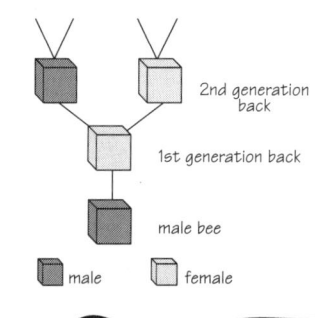

- Making sure to keep all ancestors of the same generation in the same horizontal row, continue the process of adding toothpicks and yellow and black cubes to trace back 8 complete generations in the male bee's family tree.

- As each generation is added, record the generation number, the number of male bees, the number of female bees, the number of bees in that generation, and the total number of ancestors. Organize your data in a chart.

- Look for patterns in your results. Predict the numbers of male and female bees in the 9th and 10th generations back. Be ready to explain your reasoning.

Thinking and Sharing

Invite one pair of students to draw their male bee family tree on the board. (Note: Students will realize very quickly that it becomes increasingly difficult to continue drawing past the fourth or fifth generation due to the number of branches.) While the tree is being drawn, construct a chart with column headings: *number of generations back, number of male bees, number of female bees, number of bees in generation,* and *total number of ancestors.* Invite students to record their data in the chart and share their findings.

Use prompts like these to promote class discussion:

- What was difficult about building the model of the ancestral tree?
- What did you notice as you added generations to your tree?
- As you were adding generations, did you notice any patterns that helped make your work easier? If so, explain.
- What patterns and/or relationships did you notice in the data you recorded?
- How did you go about making your predictions for the 10th generation?
- How many male bees did you find in the 9th and 10th generations? How many female bees? How many total? Explain your reasoning.
- What other interesting discoveries did you make?

Tell students that the pattern they see occurring in the chart (1, 1, 2, 3, 5, 8, 13, 21, ...) is a well-known sequence called the the *Fibonacci sequence,* named for the mathematician who discovered it.

On Their Own (Part 2)

What if... you examined the ancestry of a species in which both males and females have two parents? How would the patterns be different from those you found in Part 1?

- Work with a partner who is the same sex as you are. Construct your own family tree using blue Snap Cubes to represent males, and purple Snap Cubes to represent females. Here's how:
    - Begin by placing either a purple or blue Snap Cube on your work surface to represent you.

- Use toothpicks and a cube of each color to represent your mother and father.
- Continue adding toothpicks and cubes to represent grandparents, great-grandparents, and so on.
- Making sure to keep all ancestors of the same generation in the same horizontal row, continue the process of adding toothpicks and cubes to trace back 6 complete generations in your family tree.

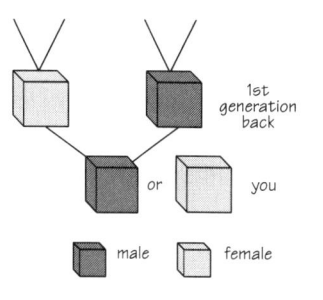

• Record your data in a chart like the one you made in Part 1.

• Look for patterns in your results. Predict the numbers of male and female ancestors in the 10th generation back. Then generalize your findings by writing algebraic expressions for the numbers of ancestors in the nth generation back. Be ready to explain your reasoning.

Thinking and Sharing

Have students help you create two class charts (one for data recorded by a female pair, and one for data recorded by a male pair) with column headings: *number of generations back*, *number of males*, *number of females*, *number of people in generation*, and *total number of ancestors*. Invite students to record their data and share their findings.

Use prompts like these to promote class discussion:

- What was difficult about building the model of your ancestral tree?
- As you were adding generations, did you notice any patterns that helped make your work easier? If so, explain.
- What patterns and/or relationships did you notice in building the tree and in the data?
- What do you notice about the data collected from trees beginning with a female as compared to those beginning with a male?
- How did you go about making your predictions for the 10th generation? for the nth generation?
- What algebraic expressions did you write to describe the number of ancestors in the nth generation?
- How might you express the total number of ancestors in n generations?

For Their Portfolio

Scientists have found many examples of the Fibonacci sequence occurring in nature. For example, the spiral patterns in pine cones, pineapples, and sunflowers, the leaf arrangements on a branch, and the number of petals on a flower are all related to the Fibonacci sequence. This sequence also appears in architecture and in other fields. Research one or more of these topics and write a short report describing how your topic is related to the Fibonacci sequence.

Teacher Talk

Where's the Mathematics?

As students work on building the family trees, they will see that the branching process spreads out very quickly and takes up considerable room. This situation provides a "built-in" motivation for students to search for patterns so that they can predict what the growth sequences will be without having to continue extending their tree.

The chart for 10 generations of the male bee is shown below.

number of generations back	number of male bees	number of female bees	number of bees in generation	total number of ancestors
original male bee	1		1	
1	0	1	1	1
2	1	1	2	3
3	1	2	3	6
4	2	3	5	11
5	3	5	8	19
6	5	8	13	32
7	8	13	21	53
8	13	21	34	87
9	21	34	55	142
10	34	55	89	231

The entries in the fourth column, 1, 1, 2, 3, 5, 8, 13, ..., form one of the most famous sequences in mathematics, the *Fibonacci sequence*. It was named for Fibonacci (Leonardo de Pisa) who lived almost 800 years ago and was the first person known to have recorded it. The sequence begins with two 1's, and each successive term is formed by adding the two previous terms. For example, 1 + 1 = 2, 1 + 2 = 3, 2 + 3 = 5, 3 + 5 = 8, and so on. Students should notice that this pattern also appears in the data for both male bees and female bees (an interesting curiosity!).

To determine the total number of ancestors in the male bee's family tree through a specific generation, students may use a cumulative addition process each time another generation is considered. Some students may notice that by finding the total number of bees two generations after the one under investigation and subtracting 2, they can find the total number of ancestors. For example, to find the total number of ancestors up to 7 generations back, subtract 2 from the entry in the *number of bees in generation* column for the ninth generation (55 – 2 = 53). Each term in the last column is 2 less than a term from the Fibonacci sequence.

The chart for the second part of the activity is shown on the following page. Students should observe that aside from the 1 that represents the student him- or herself, the data are identical whether the student is male or female.

number of generations back	number of males	number of females	number of people in generation	total number of ancestors
student	1 (if male)	1 (if female)	1	
1	1	1	2	2
2	2	2	4	6
3	4	4	8	14
4	8	8	16	30
5	16	16	32	62
6	32	32	64	126
7	64	64	128	254
8	128	128	256	510
9	256	256	512	1022
10	512	512	1024	2046
⋮	⋮	⋮	⋮	⋮
n	2^{n-1}	2^{n-1}	2^n	$2^{n+1} - 2$

The numbers appearing in the fourth column are increasing powers of 2. These form a type of sequence called an *exponential growth* sequence. For example, the number of ancestors in generation 3 is 8, which is 2 x 2 x 2, or 2^3; the number of ancestors in generation 5 is 32, which is 2 x 2 x 2 x 2 x 2, or 2^5, etc. In general, students should agree that the number of ancestors in generation n can be given by the expression 2^n, where the exponent is the same as the generation number.

Students should notice that the number of females and the number of males in each generation is the same. Each of these values is half of the total number of people in the generation and can thus be represented by the expression $\frac{2^n}{2}$, or 2^{n-1}, where n is the generation number. This expression can also be determined by recognizing that the numbers in these columns form the same pattern as those in the fourth column, with corresponding terms occurring one generation behind.

Once the number of people in each generation has been established, the total number of ancestors can be determined in several ways. Students may choose to add the number of people in each generation together, or they may notice that each entry in the last column is 2 less than the number of people in the next generation. For example, in generation 6, the total of 126 ancestors is 2 less than 128, the number of ancestors in generation 7. Since the number of people in the next generation—i.e., the (n + 1)th generation—would be 2^{n+1}, the total number of ancestors in n generations would be $2^{n+1} - 2$.

Students may be interested to learn that in his publication *Poor Richard's Almanac*, Benjamin Franklin wrote an amusing article about the number of ancestors in his own family tree. He noted that as he worked his way back through the generations, there came a time when the number of ancestors was greater than the number of people who inhabited the earth at that time! After working on their own family trees, students may have mused about this paradox themselves.

Investigating Patterns from Geometry

1. Napkins and Place Mats, page 35   (Color Tiles)
2. Pythagoras Delivers the Mail, page 40   (Geoboards)
3. Polygons, Pegs, and Patterns, page 44   (Geoboards)
4. The Airline Connection, page 48   (Circular Geoboards)
5. Inside Out, Outside In, page 52   (Pattern Blocks)

The lessons in this cluster explore patterns and functions that exist in geometric settings. The activities can be worked on in any order, and can be used within the context of either a geometry unit or a patterns/functions unit.

1. Napkins and Place Mats (Exploring patterns in quadrilaterals)

In this activity, students build sequences of squares and rectangles, and examine the patterns formed by the numbers of tiles in the borders and interiors of the quadrilaterals. As the patterns emerge, students are asked to represent them algebraically, which leads them to discover that the quantities are functions of the dimensions of the quadrilaterals.

2. Pythagoras Delivers the Mail (Investigating the Pythagorean Theorem)

This lesson serves as an introduction to the Pythagorean Theorem, exploring the relationships that exist among the areas of squares built not only on the sides of right triangles, but also on those of acute and obtuse triangles. The discussion questions help to bring to light Pythagoras's famous discovery.

In *On Their Own* Part 2, students apply what they learned about the theorem to a real-world problem situation. They then extend their thinking and their problem-solving strategies to a related problem presented in *For Their Portfolio*.

3. Polygons, Pegs, and Patterns (Exploring patterns in area)

In this activity, students make polygons on their Geoboards and try to find a relationship between the area of a polygon and the numbers of pegs in its border and its interior. In *On Their Own* Part 1, students investigate many different polygons and look for patterns. In Part 2, they try to identify a formula (Pick's Theorem) that expresses the relationship.

The activity assumes that students can find the area of Geoboard polygons. Teachers may want to precede the activity by reviewing some of the different strategies for finding area. Several of these are described in *Where's the Mathematics?* in *Pythagoras Delivers the Mail* (page 42).

4. The Airline Connection (Exploring patterns in diagonals of polygons)

In this activity, students discover that the number of diagonals that can be constructed in a polygon is a function of the number of sides in the polygon. They then apply this observation to an application problem involving airline routing.

The lesson assumes that students understand the concept of "diagonal of a polygon." *On Their Own* Part 1 also assumes an understanding of the terms *vertex* and *vertices*.

5. Inside Out, Outside In (Exploring patterns in angle measures of polygons)

This activity provides an opportunity for students to discover that the sum of the interior angles of a polygon is a function of the number of sides of the polygon, whereas the sum of the exterior angles is not.

In *On Their Own* Part 1, students must begin by finding the measures of each angle of their Pattern Blocks (without a protractor). Teachers may want to check the results of this part of the activity before allowing students to proceed.

On Their Own Part 2 assumes an understanding of the distinctions between convex and concave polygons.

NAPKINS AND PLACE MATS

- Organizing and analyzing data
- Pattern recognition
- Sequence of perfect squares
- Writing algebraic expressions

Getting Ready

What You'll Need

Color Tiles, about 60 each of 2 different colors per pair

Activity master, page 104

Overview

Students use Color Tiles to model a series of squares and rectangles. They then investigate the relationships that exist among the dimensions, the number of border tiles, and the number of interior tiles in each quadrilateral. In this activity, students have the opportunity to:

- identify and extend patterns
- explore the sequence of perfect squares
- make predictions based on patterns
- express relationships algebraically

Other *Super Source* activities that explore these and related concepts are:

Pythagoras Delivers the Mail, page 40

Polygons, Pegs, and Patterns, page 44

The Airline Connection, page 48

Inside Out, Outside In, page 52

The Activity

On Their Own (Part 1)

Nadia is preparing for a cookout and is setting the picnic table with paper napkins and place mats. She notices that the color scheme on these paper products consists of a solid color in the center surrounded by a contrasting colored border. What relationships can you find among the size of the square napkin, the area of the solid interior, and the area of the contrasting border?

- Work with a partner. Using Color Tiles, make a 3-by-3 model of a square napkin. Use one color (red, for example) to represent the solid interior, and a contrasting color (blue, for example) for the border.

- Build 3 more models of square napkins. Make each of your models one tile longer on each side than in the previous square. Again, use different colors for the interior and for the border.

- As you make each new square, record the dimensions of your model, the dimensions of the interior, the number of tiles in the interior, the number of tiles in the border, and the total number of tiles used. Organize your data in a chart.

©Cuisenaire Company of America, Inc.

- Look for patterns in your data. Predict what tiles would be needed for an 8-by-8 napkin model. Predict what tiles would be needed for a 15-by-15 napkin model. Then generalize the patterns you found by writing algebraic expressions for an *n*-by-*n* napkin model. Add these entries to your chart.

- Be ready to talk about the relationships you discovered.

Thinking and Sharing

Have students help you create a class chart displaying the data. Column headings might include: *dimensions of square napkin, dimensions of interior square, number of tiles in interior, number of tiles in border,* and *total number of tiles.*

Use prompts like these to promote class discussion:

- What patterns did you find among the entries within each column?
- What patterns and/or relationships did you find among the data in the different columns?
- What strategies did you use to find the number of tiles in the 8-by-8 square? in the 15-by-15 square?
- What tiles would be needed for the *n*-by-*n* square?
- What algebraic expressions did you write to express the relationships you found?

On Their Own (Part 2)

What if... the rectangular place mats, like the napkins, have a solid interior surrounded by a contrasting colored border? Will the numerical relationships formed by this design be the same as those you found for the square napkins?

- Using Color Tiles, make a 3-by-5 model of a rectangular place mat. Use two different colors, one for the interior and another for the border.

- Build 3 more models of place mats, increasing the width by 1 tile and the length by 2 tiles for each subsequent rectangle. Again, use different colors for the interior and for the border.

- As you make each new rectangle, record its dimensions, the dimensions of the interior, the number of tiles in the interior, the number of tiles in the border, and the total number of tiles used. Organize your data in a chart.

- Look for patterns in your data. Predict what tiles would be needed for an 8-by-15 place mat model. Predict what tiles would be needed for a 15-by-29 place mat model. Add these entries to your chart.

- For each place mat model, look to see if you can find a relationship between its width and the rest of its data. Then generalize the relationships you find by writing algebraic expressions to represent the data for a place mat whose width is *n* units.

- Be ready to explain your work.

Thinking and Sharing

Have students help you create a class chart similar to the one from the first activity. Encourage students to make comparisons between the data on the two charts.

Use prompts like these to promote class discussion:

- What patterns did you find among the entries within each column?
- What patterns and/or relationships did you find among the data in the different columns?
- What strategies did you use to find the number of tiles in the 8-by-15 rectangle? in the 15-by-29 rectangle?
- What tiles would be needed for a rectangle with width n units?
- What algebraic expressions did you write to express the relationships you found?
- Are the interior rectangles similar to the full rectangles? How do you know?

For Their Portfolio

Explain the methods you used in looking for patterns, analyzing the data, and writing algebraic expressions to generalize your findings.

Teacher Talk

Where's the Mathematics?

Students will discover a number of patterns emerging from the data they collect. The data for the first activity is shown below.

dimensions of square napkin	dimensions of interior square	number of tiles in interior	number of tiles in border	total number of tiles in square
3 x 3	1 x 1	1	8	9
4 x 4	2 x 2	4	12	16
5 x 5	3 x 3	9	16	25
6 x 6	4 x 4	16	20	36
7 x 7	5 x 5	25	24	49
8 x 8	6 x 6	36	28	64
9 x 9	7 x 7	49	32	81
10 x 10	8 x 8	64	36	100
⋮	⋮	⋮	⋮	⋮
15 x 15	13 x 13	169	56	225
⋮	⋮		⋮	⋮
n x n	$(n-2)$ x $(n-2)$	$(n-2)$ x $(n-2)$,	$4n-4$	n x n, or n^2

©Cuisenaire Company of America, Inc.   NAPKINS AND PLACE MATS ♦ Patterns/Functions ♦ Grades 7–8

Students should recognize that the total number of tiles needed for each square is the product of its dimensions. In general, if a square measures n units on a side, its area is $n \times n$, or n^2. Students may also notice that these products (appearing in both the third and fifth columns of the chart) increase by consecutive odd integers. For example, $16 - 9 = 7$, $25 - 16 = 9$, $36 - 25 = 11$, and so on.

To find a way to generalize the number of tiles in the interior, students need to recognize that the interior is a square whose side lengths are always 2 tiles shorter than the side lengths of the whole square. For example, in an 8-by-8 square napkin, the inner square measures 6 by 6. Thus, if the square napkin measures n tiles on a side, then the interior square will measure $(n - 2)$ tiles on a side and have an area of $(n - 2) \times (n - 2)$, or $(n - 2)^2$, square units. The sequence generated by these areas (1, 4, 9, 16, 25, ...) is called the *sequence of perfect squares*. As mentioned above, the differences between consecutive terms in this sequence increase by consecutive odd integers.

As they build successively larger models, students may notice that the number of tiles in each border is 4 more than the number of tiles in the previous border. They should also recognize that each of these numbers, in addition to being a multiple of 4, is also the difference between the total number of tiles in the whole napkin and the number of tiles in the interior, each a perfect square.

In attempting to define the relationship between the length of the side of a square and the number of tiles in its border, some students may reason (although incorrectly) that if the side of the square is composed of n tiles, the border must contain $4n$ tiles (thinking of the border as the perimeter of the square). When they try to verify this expression using the data, they will see that it does not work. They may then realize that they must take into account that each of the 4 corner tiles contributes 2 units to the perimeter but needs to be counted only once in the tile count. Subtracting 4 from the perimeter eliminates this duplication. Thus, the number of border tiles can be represented by the algebraic expression $4n - 4$.

The data for the second activity is shown below. Many of the patterns for the square (discussed above) can be modified to generate patterns for the rectangle.

dimensions of rectangular place mat	dimensions of interior rectangle	number of tiles in interior	number of tiles in border	total number of tiles in rectangle
3 x 5	1 x 3	3	12	15
4 x 7	2 x 5	10	18	28
5 x 9	3 x 7	21	24	45
6 x 11	4 x 9	36	30	66
7 x 13	5 x 11	55	36	91
8 x 15	6 x 13	78	42	120
9 x 17	7 x 15	105	48	153
10 x 19	8 x 17	136	54	190
⋮	⋮	⋮	⋮	⋮
15 x 29	13 x 27	351	84	435
⋮	⋮	⋮	⋮	⋮
$(n) \times (2n - 1)$	$(n - 2) \times (2n - 3)$	$(n - 2)(2n - 3)$	$2(n) + 2(2n - 1) - 4$, or $6n - 6$	$n(2n - 1)$

Students may begin by looking at the relationship between the length of the rectangle and its width. As students compare the two dimensions, they may discover that the length is always 1 less than twice the width. For example, when the width is 4, the length is 1 less than 2 x 4, or 7; when the width is 9, the length is 1 less than 2 x 9, or 17. In general, the width and length of the rectangle can be represented by n and $2n - 1$.

Like the square napkin, the rectangular place mat has a smaller quadrilateral in its interior whose dimensions are each 2 less than the larger, outer quadrilateral. For example, the dimensions of the interior rectangle in a 6-by-11 place mat are 4 by 9. In general, a rectangle with dimensions n by $(2n - 1)$ will contain an interior rectangle with dimensions $(n - 2)$ by $[(2n - 1) - 2]$, or $(n - 2)$ by $(2n - 3)$. Students may arrive at these same dimensions by noticing that the length of the interior rectangle is 1 more than twice its width.

Note that the rectangle formed by the place mat is not similar to the interior rectangle. Students may verify this by comparing the ratios of corresponding sides.

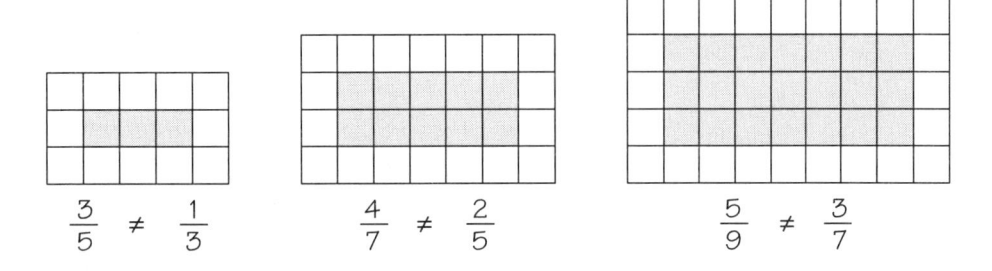

$$\frac{3}{5} \neq \frac{1}{3} \qquad \frac{4}{7} \neq \frac{2}{5} \qquad \frac{5}{9} \neq \frac{3}{7}$$

Students may find it interesting to discover that the data in the *number of tiles in the interior* column and the data in the *total number of tiles in the rectangle* column form sequences in which the terms increase by every other odd integer. For example, in the last column, 28 – 15 = 13, 45 – 28 = 17, 66 – 45 = 21, and so on. In terms of the number of tiles in the border of each rectangle, students can apply what they learned from their work with squares, subtracting 4 from the expression representing the perimeter of the rectangle.

PYTHAGORAS DELIVERS THE MAIL

- Properties of triangles
- Pythagorean Theorem
- Area

Getting Ready

What You'll Need

Geoboards, 1 per student

Rubber bands

Rulers

Dot paper, page 119

Activity master, page 105

Overview

Students examine squares built on the sides of right, obtuse, and acute Geoboard triangles. They look for relationships between the areas of the squares and the type of triangle on which they are built. In this activity, students have the opportunity to:

- devise methods for finding areas
- formulate and test generalizations
- learn about and apply the Pythagorean Theorem
- use mathematical reasoning to solve a real-world problem

Other *Super Source* activities that explore these and related concepts are:

Napkins and Place Mats, page 35

Polygons, Pegs, and Patterns, page 44

The Airline Connection, page 48

Inside Out, Outside In, page 52

The Activity

On Their Own (Part 1)

Pythagoras, one of the world's most famous mathematicians, discovered interesting relationships between types of triangles and the lengths of their sides. Can you retrace the steps leading to his remarkable discovery?

- Work with a group of at least 4 students. Each of you should make a different-sized right triangle on your Geoboard.
- Record your triangle on dot paper near the center of the paper.
- Using a ruler, draw a square on each side of your triangle. Make the sides of each square congruent to the side of the triangle on which it is built.
- Find and record the area of each of your three squares. Let the area of one small dot-paper square be the unit of measure.
- Discuss and check the work of the other members of your group.
- Repeat the activity using obtuse triangles and then acute triangles.
- Look for relationships among the areas of the three squares surrounding each type of triangle. Generalize the findings of your group.

Thinking and Sharing

Invite groups to share their recordings and their observations about the areas of the squares they built on the sides of the different types of triangles. Encourage them to help each other find the areas of squares they may have had difficulty working with.

Use prompts like these to promote class discussion:

- What method(s) did you use to find the areas of your squares?
- For which squares was it hard to find the area? For which was it easy? Why?
- What patterns did you discover?
- What generalizations can you make about the areas of the three squares built on the sides of a right triangle? an obtuse triangle? an acute triangle?
- Do you think your findings would be consistent for other triangles of the same type? How do you know?

You may want to tell students that the relationship that holds for right triangles (the sum of the areas of the squares on the legs is equal to the area of the square on the hypotenuse) is called the *Pythagorean Theorem*, discovered by and named for Pythagoras.

On Their Own (Part 2)

> What if... a letter measuring 1/8" by 11" by 14" is sent to a teacher who has an open-front rectangular mailbox in the school office measuring 5" high by 10" wide by 15" deep. Will the letter fit into the teacher's mailbox without being bent or folded?
>
> - Draw a model of the front of the mailbox on dot paper.
> - Discuss with your group how the letter might be positioned in the mailbox, and sketch it on your dot paper drawing.
> - Using what you learned in the first activity, decide whether the mailbox can accommodate the teacher's letter. Be ready to justify your conclusions.

Thinking and Sharing

Invite groups to share their conclusions and tell how they reached them.

Use prompts like these to promote class discussion:

- What information was needed to draw the mailbox model on dot paper?
- How did you choose to position the letter in the mailbox? Why?
- How did you apply what you learned in the first activity to solve this problem?
- What did you conclude about whether the letter would fit? Explain your reasoning.
- What size restrictions would you place on any piece of mail if it is to fit in the mailbox?

For Their Portfolio

Write a paragraph or two describing how you might approach the mailbox problem if the mail was a package instead of a letter.

Teacher Talk

Where's the Mathematics?

Students should be familiar with the concepts of *right triangles* (those containing a 90° angle), *acute triangles* (those in which each angle measures less that 90°), and *obtuse triangles* (those containing an angle whose measure is greater than 90°). Working in groups of at least four, students should be able to build and investigate a good number of each of these types of triangles.

As students build squares on the sides of their triangles, they will discover that some of the squares are more easily constructed than others. They also may use different strategies to find the areas of their squares. For those squares that have sides parallel to the edges of the dot paper, some students may simply count the number of unit squares contained in the squares. Other students may recognize that the areas of these squares can be found by multiplying the length of one side of the square by itself.

To find the area of a square whose sides are not parallel to the edges of the dot paper, some students may partition the square into shapes whose areas may be easier to find. Others may enclose them inside larger squares whose sides *are* parallel to the edges of the dot paper. The areas of these "enclosed squares" can then be determined by finding the area of the larger, surrounding square and subtracting the areas of the four corner triangular areas. Examples of these methods are shown at the right.

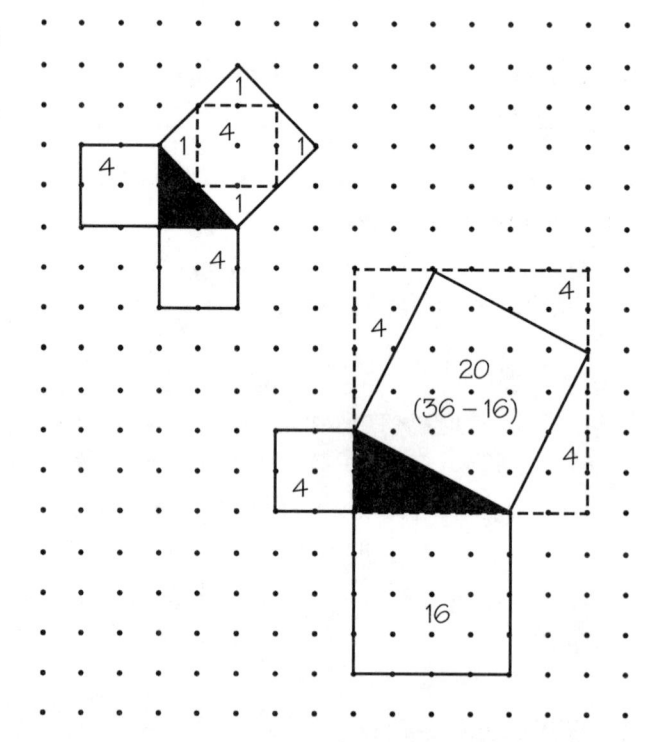

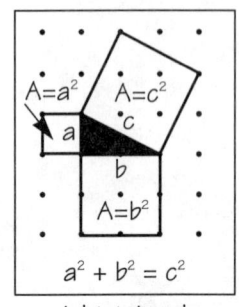

In examining the areas of the squares surrounding each right triangle, students should notice that the sum of the areas of the squares on the two shorter sides (the legs) is equal to the area of the square on the longest side (the hypotenuse). This relationship is known as the *Pythagorean Theorem* and can be summarized as follows: In a right triangle, if a and b are the lengths of the legs and c is the length of the hypotenuse, then $a^2 + b^2 = c^2$.

$A = a^2$ $A = c^2$
$A = b^2$
$a^2 + b^2 = c^2$

right triangle

For obtuse triangles, students will find that the area of the square on the longest side of the triangle is greater than the sum of the areas of the two smaller squares. For acute triangles, students should find that the area of the square on the longest side of the triangle is less than the sum of the areas of the two smaller squares. If the acute triangle is isosceles or equilateral, there may not be a "longest side," but the same inequality will exist based on the relative lengths of the sides.

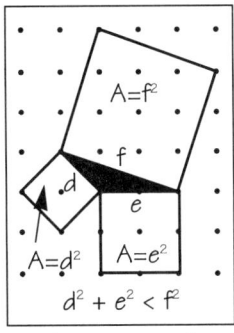

obtuse triangle

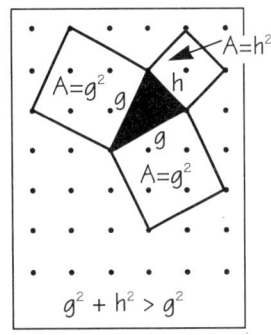

acute isosceles triangle

Part 2 of the activity gives students opportunities to work on a practical application of the Pythagorean Theorem and to build a mathematical model that can be used as an aid in analyzing the problem. Students may realize that the length of the diagonal of their dot paper rectangle is the width of the widest piece of mail that will fit into the mailbox. It is also the hypotenuse of two congruent right triangles.

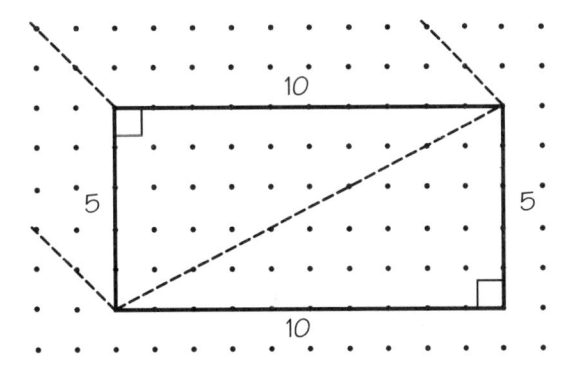

By applying the Pythagorean Theorem to either one of the right triangles, students can find the area of the square that can be built on the diagonal ($5^2 + 10^2 = 125$ square units), and then consider the length of the diagonal itself, keeping in mind that the diagonal is the side of this square. They may estimate that the side length is between 11 and 12 units. Students who use a calculator may take the square root of 125 to arrive at the decimal approximation 11.18. Either way, students can conclude that the 11" width of the letter can be accommodated in the mailbox.

The depth of the mailbox might be considered extraneous information in this problem because students might reason that a piece of mail could extend past the opening of the mailbox. Also, if the piece of mail was significantly thicker than 1/8", the letter might need to be less than 11" wide to fit in the mailbox, as the thickness would prevent it from fitting into the corners.

POLYGONS, PEGS, AND PATTERNS

- Area
- Using patterns
- Writing formulas
- Pick's Theorem

Getting Ready

What You'll Need

Geoboards, 1 per student

Rubber bands

Geodot paper, page 117

Activity master, page 106

Overview

Students create a variety of polygons on their Geoboards, each having specified numbers of boundary pegs and interior pegs. They then find the areas of their polygons and search for patterns in their data. In this activity, students have the opportunity to:

- collect and organize data
- look for patterns and use them to make predictions
- experiment with writing and testing algebraic formulas
- discover Pick's Theorem
- reinforce strategies for finding areas

Other *Super Source* activities that explore these and related concepts are:

Napkins and Place Mats, page 35

Pythagoras Delivers the Mail, page 40

The Airline Connection, page 48

Inside Out, Outside In, page 52

The Activity

On Their Own (Part 1)

> Did you ever wonder whether the area of a Geoboard polygon might be related to the number of pegs on its boundary and in its interior? What patterns can you find that might prove this to be true?
>
> - Work in a group. Using your Geoboard, explore all possible areas for polygons that contain only 3 boundary pegs. Start by investigating polygons that have 3 boundary pegs and no interior pegs, then 3 boundary pegs and 1 interior peg, 3 boundary pegs and 2 interior pegs, and so on.
> - Copy each polygon onto geodot paper and record the number of boundary pegs, the number of interior pegs, and the area.
> - Now investigate polygons having 4, 5, and 6 boundary pegs in the same way.
> - Organize your findings in a chart. Look for patterns in the data that can be used to help predict the areas of other polygons.

Thinking and Sharing

Have students post several examples of each kind of polygon on the board. Ask them to share their observations, point out the patterns they discovered, and discuss any generalizations they made. Encourage other students to test the generalizations on their shapes. Have groups share their charts and discuss different ways of organizing the data to facilitate pattern recognition.

Use prompts like these to promote class discussion:

- Which polygons had areas that were easy to determine? Which polygons had areas that were more difficult to determine? Why?
- What strategies did you use to find the areas?
- What patterns did you find in your data? Do you think the patterns will work for other polygons? How do you know?
- How could the patterns be used to find the areas of other Geoboard polygons?
- What would be the area of a polygon with 7 boundary pegs and no interior pegs? What about one with 8 boundary pegs and 1 interior peg? Explain your reasoning.

On Their Own (Part 2)

What if... you wanted to write a formula for the area of a polygon based upon the numbers of its boundary pegs and interior pegs? How might you do this?

- Look at the data you collected in the first part of the activity. See if you can figure out how the area of a shape could be calculated using the number of boundary pegs (B) and the number of interior pegs (I). Consider simple algebraic expressions containing various operations.

- If you find a formula that seems to work for a particular set of data, test it on other sets. If it doesn't work, try modifying it in some way and then test it again.

- If you have trouble finding a formula that works for all polygons, consider the patterns formed by the data and the types of numbers that appear in the different columns of your charts. Consider the answers to questions such as "When is the area of the polygon a whole number of units? When is it a half number of units?" and "What is the relationship between the area and the number of boundary pegs when there are no interior pegs?"

- Share your ideas with other groups. If you discover a formula that works for polygons with 3, 4, 5, and 6 boundary pegs, test it on Geoboard polygons having 7 or 8 boundary pegs.

- Be ready to share your findings and to talk about the strategies you used in attempting to derive a formula.

Thinking and Sharing

Invite students to share their formulas and to talk about the work they did in attempting to derive them. Encourage students who were unsuccessful in finding a formula to join in the discussion and to tell about the algebraic expressions they investigated.

Use prompts like these to promote class discussion:

- What relationships did you find that helped you get started?
- What algebraic expressions did you try? What happened?
- Did you find any patterns in the data that provided clues to the formula? If so, tell about them and explain how they helped you write your formula.
- Did analyzing the types of numbers that appear in the data help you to make any discoveries that proved useful? If so, explain.
- If you were successful in finding a formula that worked, how did you "zero in" on it?

Explain to students that the formula that relates these numbers is called *Pick's Theorem*, named after the mathematician who discovered it. Some groups may have more difficulty than others in deriving the formula. You may want to provide a little extra guidance for these groups (see *Where's the Mathematics?*).

Note that there is much to be gained from the search for the formula, even when the desired result is not obtained. Students have the opportunity to write and evaluate meaningful algebraic expressions and to work together in a collaborative format to organize and analyze data from their own investigation.

For Their Portfolio

Describe the techniques you used in trying to derive your formula. Explain how you went about analyzing the data and formulating ideas to try, and what happened as your investigation proceeded.

Teacher Talk

Where's the Mathematics?

In this activity, students learn that being able to collect, sort, and organize information is essential when learning how to analyze data and draw conclusions. In the first part of the activity, students may be able to predict the areas of their polygons based on the patterns they notice without knowing specific formulas. Although they may organize their data in different ways, their investigations should yield the following results:

boundary pegs	interior pegs	area in sq. units
3	0	½
3	1	1½
3	2	2½
3	3	3½
3	4	4½
3	5	5½
3	6	6½

boundary pegs	interior pegs	area in sq. units
4	0	1
4	1	2
4	2	3
4	3	4
4	4	5
4	5	6
4	6	7

boundary pegs	interior pegs	area in sq. units
5	0	1½
5	1	2½
5	2	3½
5	3	4½
5	4	5½
5	5	6½
5	6	7½

boundary pegs	interior pegs	area in sq. units
6	0	2
6	1	3
6	2	4
6	3	5
6	4	6
6	5	7
6	6	8

Students should discover that shapes with the same number of boundary pegs and the same number of interior pegs have the same area. They may also notice that when the number of interior pegs remains constant and the number of boundary pegs is increased by 1, the area of the polygon is increased by ½ square unit. Therefore, by looking at the areas of the Geoboard polygons with 3, 4, 5, or 6 boundary pegs and no interior pegs, students should be able to conclude that a Geoboard polygon with 7 boundary pegs and no interior pegs will have an area of 2½ square units. Similarly, a Geoboard polygon with 8 boundary pegs and 1 interior peg will have an area of 4 square units.

In searching for the formula relating the area of a Geoboard polygon to the numbers of its boundary and interior pegs, students may consider the types of numbers that appear in their data. The numbers of pegs are, of course, whole numbers, while the areas of the shapes are sometimes whole numbers and sometimes mixed numerals with fractions of ½. Some students may reason that this would suggest that one of the peg counts must be divided by 2 (or multiplied by ½) in the formula. By identifying those polygons whose area is a whole number as those that have an even number of boundary pegs, and those polygons whose area is a mixed number as those that have an odd number of boundary pegs, students may realize that the boundary peg count is the critical piece of data in generating the appropriate type of number for the area. Taking half of an even number yields a whole number, while taking half of an odd number yields a mixed numeral whose fraction is ½.

Students may also notice that when the number of boundary pegs remains constant and the number of interior pegs increases by 1, the area of the polygon also increases by 1 square unit.

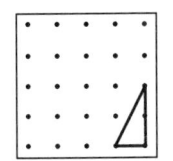

4 boundary pegs
no interior pegs
Area = 1 square unit

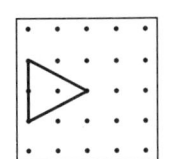

4 boundary pegs
1 interior peg
Area = 2 square units

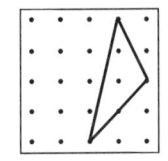

4 boundary pegs
2 interior pegs
Area = 3 square units

With these ideas in mind, students may experiment with different algebraic expressions involving one-half of the number of boundary pegs (B) and the number of interior pegs (I). Through trial and error, students may discover that when adding $B/2$ and I, the total is always 1 more than the area. Thus, by subtracting 1 from this sum, the desired result can be obtained. This relationship is called *Pick's Theorem* and can be written algebraically as: Area = $B/2 + I - 1$, where B represents the number of boundary pegs and I represents the number of interior pegs of the Geoboard polygon.

THE AIRLINE CONNECTION

- Spatial visualization
- Pattern recognition
- Writing formulas
- Properties of polygons

Getting Ready

What You'll Need

Circular Geoboards, 1 per student

Rubber bands

Circular geodot paper, page 118

Activity master, page 107

Overview

Using circular Geoboards, students create polygons, make their diagonals, and look for a way to relate the number of diagonals that can be made in a polygon to the number of sides in the polygon. They then apply their discoveries to an application involving airline routing. In this activity, students have the opportunity to:

- collect and organize data
- search for patterns and formulate generalizations
- write and test algebraic formulas
- apply mathematical concepts to a real-world situation

Other *Super Source* activities that explore these and related concepts are:

Napkins and Place Mats, page 35

Pythagoras Delivers the Mail, page 40

Polygons, Pegs, and Patterns, page 44

Inside Out, Outside In, page 52

The Activity

On Their Own (Part 1)

Polygons and their diagonals have many real-world applications, one of which you'll investigate in Part 2 of this activity. But first, can you find a way to predict the number of diagonals that can be made in a polygon?

- Work with a partner. Each of you should make a polygon on the circular side of a Geoboard. Make polygons whose vertices are located at pegs on the circle. Do not use the peg at the center of the circle.
- Pick a vertex and make as many diagonals as you can from that vertex. Repeat this process from each vertex of your polygon.
- Draw your polygon and its diagonals on circular geodot paper. Record the number of sides in your polygon, the number of vertices, the number of diagonals from each vertex, and the total number of diagonals.
- Repeat this process, each time making a polygon with a different number of sides.

- Organize your data with those of your partner and create a chart presenting your findings. Look for patterns in your data.
- Predict the number of diagonals that can be made in a polygon with *n* sides. Write a formula that can be used to generalize your conclusion.

Thinking and Sharing

You may want to have students cut out their drawings and post them for the class to see. If so, group students' work according to the number of sides of their polygons (3-sided polygons, 4-sided polygons, 5-sided polygons, and so on). Have them help create a class chart, adding the data for each new class of polygons as it is posted.

Use prompts like these to promote class discussion:

- What do you notice about the posted polygons?
- What is the smallest number of diagonals a polygon had? the greatest number?
- Could you predict how many diagonals a polygon had before you made them all? If so, how did you do this?
- What patterns and relationships did you discover?
- How can you determine the number of diagonals a polygon has if you know the number of sides it has?
- What formula did you write to generalize the relationship you found?

On Their Own (Part 2)

What if... Alpha Airlines has just been granted permission to service 8 new cities in the United States and must hire more pilots? Each pilot will fly direct roundtrip flights between 2 of the 8 cities. If there must be direct flights between every pair of cities, how many pilots must be hired by Alpha Airlines?

- Discuss with your partner how you might use the circular Geoboard and what you learned in Part 1 to solve Alpha's hiring problem. You may want to begin by using your Geoboard to construct a model of the airline's routing plan.
- As you work with your model, you may want to consider the following questions: How many flights will originate from one city? How is this problem similar to the problem in Part 1? How is it different? How do the differences affect the problem?
- Using your model, determine the number of pilots that Alpha must hire.
- Write an algebraic expression that could be used to determine the number of pilots who need to be hired if the airline is to service *n* new cities. Be ready to justify your reasoning.

Thinking and Sharing

Invite students to share their strategies for solving the problem. Encourage them to also explain their thought process in generalizing their findings.

Use prompts like these to promote class discussion:

- How did you use your circular Geoboard to model the problem?
- How is this problem similar to the first activity? How is it different?
- How did knowing the formula for the number of diagonals in a convex polygon help in solving the airline problem?
- How did you figure out the number of pilots needed to service n new cities?
- How can you generalize what you've discovered to make a statement about line segments and points?

For Their Portfolio

Write a paragraph or two describing a real-world situation that can be modeled by a polygon and its diagonals. Try to express the situation in terms of a problem that needs to be solved.

Teacher Talk

Where's the Mathematics?

Ten different types of polygons can be made on the circular Geoboard, ranging from those with 3 sides to those with 12 sides. The shapes students make will vary considerably depending on the pegs that they choose to use as vertices.

The chart for the polygons and their diagonals is shown below.

number of sides	name of polygon	number of vertices	number of diagonals from 1 vertex	total number of diagonals
3	triangle	3	0	0
4	quadrilateral	4	1	2
5	pentagon	5	2	5
6	hexagon	6	3	9
7	heptagon	7	4	14
8	octagon	8	5	20
9	nonagon	9	6	27
10	decagon	10	7	35
11	11-gon	11	8	44
12	dodecagon	12	9	54
⋮	⋮	⋮	⋮	⋮
n	n-gon	n	$n-3$	$\dfrac{n(n-3)}{2}$

Students should notice that the number of vertices is always the same as the number of sides of the polygon. They also may notice that the number of diagonals from a single vertex is always three less than the number of sides (or vertices) of the polygon. This pattern occurs because once a particular vertex has been chosen, diagonals can be formed by connecting that vertex to all but three vertices – itself, and the two adjacent vertices of the polygon. In general, if a polygon has n sides (or vertices), $(n-3)$ diagonals can be constructed from each vertex.

Students might reason that by multiplying the number of diagonals originating from each vertex by the number of vertices of the polygon, the total number of diagonals for the shape can be determined. This is partially correct, but students need to recognize that, in doing this, they've counted each diagonal twice. For example, a diagonal constructed from vertex A to vertex B is the same diagonal as the one constructed from vertex B to vertex A. To account for this, the product must be divided by 2. In general, the number of diagonals, d, that can be constructed in a convex polygon with n sides, can be found using the formula $d = \dfrac{n(n-3)}{2}$.

To solve the airline problem, students can represent each of the 8 cities in the problem by a peg on the circular Geoboard. A roundtrip flight can then be represented by a rubber band joining 2 pegs. Using this model, students can determine that a pilot would need to be hired for each rubber band they place on the Geoboard.

The difference between the problems presented in Part 1 and Part 2 lies in the fact that in Part 2, each city must have a direct flight going to each of the other cities. (In constructing diagonals of a polygon in Part 1, students only connected vertices that were non-adjacent.) Thus, with 8 new cities to service, there would be $(8-1)$, or 7 direct flights (line segments) from each city (vertex). If there are n cities to service, $(n-1)$ flights would be needed to connect one city to all other cities.

Students may use a process similar to the one they used in Part 1 to calculate the total number of pilots needed for the new routes. By multiplying the number of routes from each city (7) by the number of cities (8), and dividing by 2 to account for the fact that each pilot would be flying round trips, students should find that 28 pilots would need to be hired. In general, if n cities are added to the airline service, $\dfrac{n(n-1)}{2}$ new pilots would be needed.

Some students may realize that their model of the airline problem is an octagon and its diagonals. They may also recognize that the solution to the problem involves counting not only the diagonals of their octagon, but also its sides. From their work in Part 1, students know that the total number of diagonals that can be constructed in an octagon is 20. Adding the 8 sides of the octagon gives a total of 28 line segments, each of which, in the airline problem, represents a pilot that needs to be hired. In the general case, adding the number of diagonals, $\dfrac{n(n-3)}{2}$, to the number of sides, n, produces an expression equivalent to the expression above.

$$\dfrac{n(n-3)}{2} + n$$

$$\dfrac{n(n-3)}{2} + \dfrac{2n}{2}$$

$$\dfrac{n^2 - 3n + 2n}{2}$$

$$\dfrac{n^2 - n}{2} = \dfrac{n(n-1)}{2}$$

INSIDE OUT, OUTSIDE IN

- Pattern recognition
- Convex and concave polygons
- Angles of polygons
- Writing formulas

Getting Ready

What You'll Need
Pattern Blocks, several of each shape per student

Rulers

Activity master, page 108

Overview
Using Pattern Blocks, students investigate the sums of the measures of the interior angles and of the exterior angles of a variety of polygons. In this activity, students have the opportunity to:

- build polygons containing various types of angles
- develop strategies for measuring angles
- distinguish between convex and concave polygons
- write and evaluate algebraic expressions

Other *Super Source* activities that explore these and related concepts are:

Napkins and Place Mats, page 35

Pythagoras Delivers the Mail, page 40

Polygons, Pegs, and Patterns, page 44

The Airline Connection, page 48

The Activity

On Their Own (Part 1)

> Angela has a hypothesis that the sum of the interior angle measures of a polygon is related to the number of sides in the polygon. Can you help Angela test her theory?
>
> - Working with a partner, find the measure of each angle of each Pattern Block shape. Record your findings by tracing the Pattern Blocks on paper and labeling each angle with its measure. Also record the sum of the angle measures for each shape.
> - Make new polygons by combining Pattern Blocks in various ways.
> - Copy your polygons onto paper. For each polygon, find the measure of each of its interior angles. Record these measures on your drawings. Also record the number of sides of the polygon, and the sum of its angle measures.
> - Organize your findings and look for patterns.
> - Use your data to predict the sum of the interior angles of a polygon with 12 sides. Then write an algebraic expression that can be used to find the sum of the interior angle measures of a polygon with *n* sides.

Thinking and Sharing

Invite a pair of students to display one of its combination shapes and tell about the methods they used to find each angle measure. Have them record the number of sides of their shape and the sum of its angle measures on the chalkboard. Ask other students who built polygons with the same number of sides to tell how their data compare with this information. Repeat this for polygons having different numbers of sides. Then have students help you organize the results into a class chart, listing the polygons from those having the fewest sides to those having the most.

Use prompts like these to promote class discussion:

- How did you go about finding the measures of the angles of your polygon?
- Were some angle measures easier or harder to find than others? Which ones? Why?
- How would you describe the differences and/or similarities in the shapes of the polygons?
- What patterns did you discover?
- What predictions did you make about the sum of the angle measures of a polygon with 12 sides? with n sides?

On Their Own (Part 2)

What if... you found the measures of the <u>exterior</u> angles of different polygons? Do you think you might find a relationship between the sum of their measures and the number of sides in the polygon?

- For each convex shape you recorded in the first activity, use a ruler to extend one of the sides of the polygon at each vertex to form exterior angles.
- Determine and record the measure of each exterior angle of your polygons. Also record the number of sides of each polygon and and the sum of its exterior angle measures.
- Organize your findings and look for patterns.
- Use your data to predict the sum of the measures of the exterior angles of a convex polygon with 12 sides. Then write a general statement about the sum of the exterior angle measures of any convex polygon.

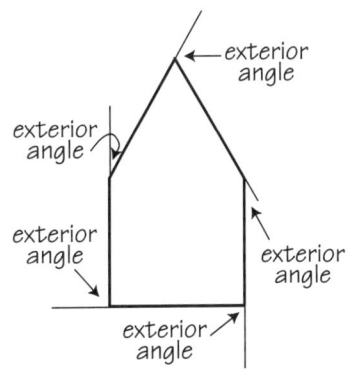

Thinking and Sharing

Encourage students to share their methods for finding the measures of exterior angles. Ask them also to tell about any observations they made during their investigation.

Use prompts like these to promote class discussion:

- How did you go about finding the measures of the exterior angles of your convex polygons?
- Were some angle measures easier or harder to find than others? Which ones? Why?
- What relationship exists between each interior angle and its adjacent exterior angle?
- What patterns did you discover?
- What predictions could you make about the sum of the exterior angle measures of a convex polygon with 12 sides? with n sides?

For Their Portfolio

Suppose you know the sum of the interior angles of a particular *regular* polygon (a polygon that is both equilateral and equiangular). Explain how you might figure out the number of sides in the polygon and the measure of each interior and each exterior angle.

Teacher Talk

Where's the Mathematics?

Students will need to determine the measures of the angles of each Pattern Block shape before they start combining blocks to create larger shapes. Their measures are shown below.

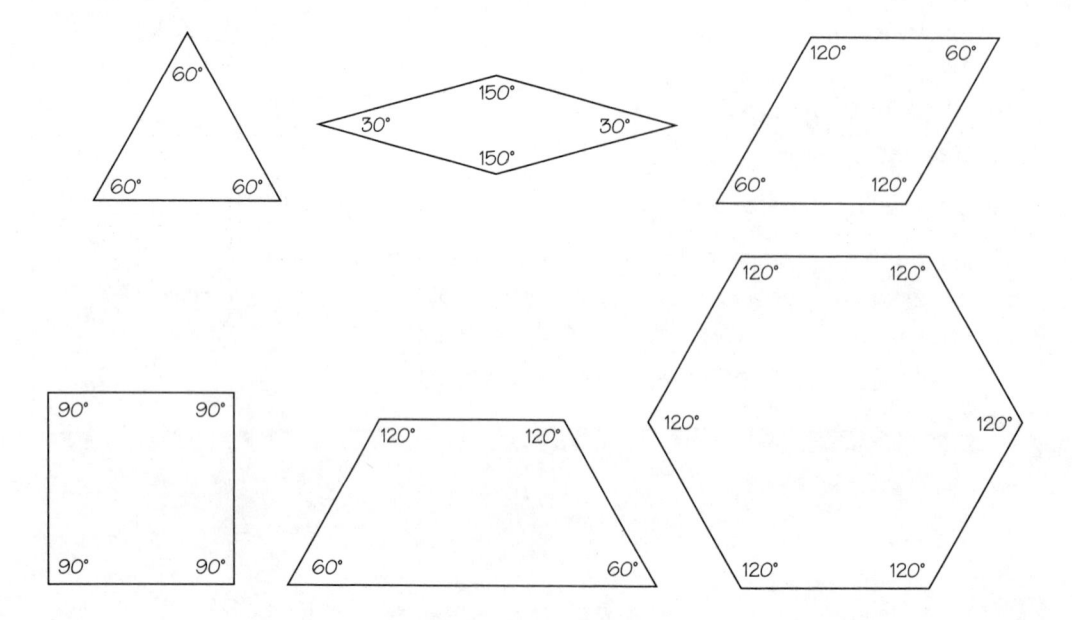

54   the Super Source • Patterns/Functions • Grades 7-8   ©Cuisenaire Company of America, Inc.

As students work with the Pattern Blocks to build different-shaped polygons, they may notice differences in appearance among their polygons. Some of their polygons, those that are convex, will have angles that all measure less than 180°, while others, those that are concave, will have one or more angles that measure more than 180°.

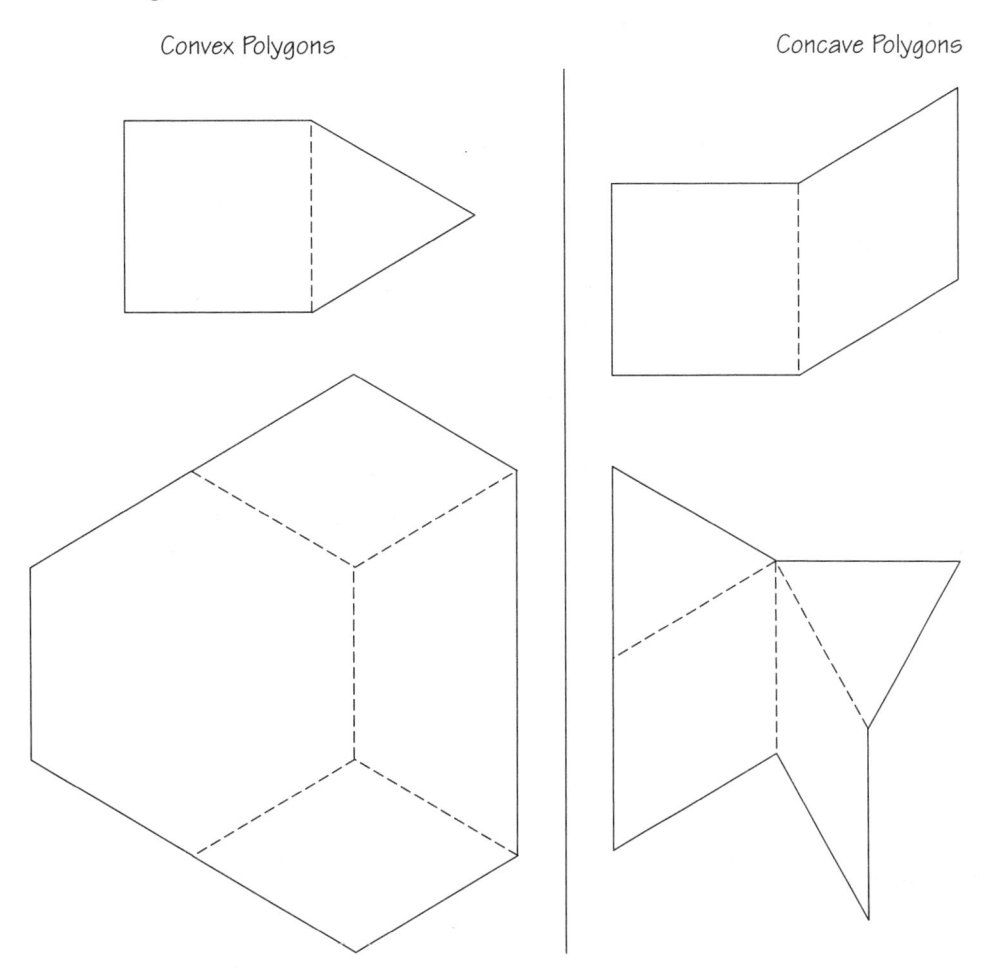

Convex Polygons | Concave Polygons

Some students may have more difficulty measuring the angles of concave polygons than of convex polygons. To measure the angles that are greater than 180°, some students may simply add the measures of the angles of the Pattern Blocks that were used to form the angle. Other students may find the measure of the smaller angle in the exterior and subtract it from 360° (the measure of one rotation) to find the measure of the larger, interior angle.

As class members share their results, students will discover that for polygons with the same number of sides, the sum of the measures of the interior angles is the same. This is true regardless of whether the polygon is convex or concave.

number of sides in polygon	sum of the interior angle measures
3	180°
4	360°
5	540°
6	720°
7	900°
8	1080°
9	1260°
10	1440°
11	1620°
12	1800°
⋮	⋮
n	$180°(n-2)$

Students should observe that the sums of the interior angle measures increase by 180° as the number of sides in the polygon increase by 1. By expressing the sums in the form (180° + 180° + 180° + ...), students may discover that the number of addends of 180° is always 2 less than the number of sides in the polygon. For example, in a polygon with 7 sides, the sum of the interior angle measures is 180° + 180° + 180° + 180° + 180°, or 900°. To find the sum of the interior angle measures of a polygon with 12 sides, students can add 10 addends of 180° (or multiply 180° by 10) to obtain 1800°. Students should generalize that the sum of the interior angle measures of a polygon with n sides is $(n - 2) \times 180°$.

Students may use a variety of approaches to find the measures of the exterior angles in Part 2 of the activity. Some students may measure these angles using their Pattern Blocks. Other students may notice that each exterior angle and its adjacent interior angle are supplementary; that is, their measures total 180°. By subtracting each known interior angle measure from 180°, the measure of the exterior angle can be found. Students can think about this method as working from the "inside out."

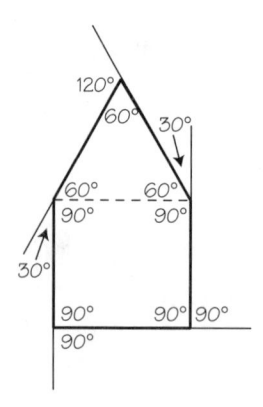

$$120° + 30° + 90° + 90° + 30° = 360°$$

When students determine the measures of the exterior angles of each polygon and find their total, they should discover that the sum is 360°, regardless of the number of sides in the original convex polygon. Students should be able to summarize this relationship in a statement such as "In any convex polygon, the sum of the measures of the exterior angles is always 360°."

You may want to point out that if the measures of the exterior angles of a convex polygon are known, students can work from the "outside in" to find the measures of the interior angles. These measures could be determined using the supplementary angle relationship and then added to find the sum.

Investigating Patterns and Functions as Problem-Solving Tools

1. Count Square and Countess Triangle, page 58   (Color Tiles and Pattern Blocks)
2. Greek Border Designs, page 63   (Color Tiles)
3. Table for 63, Please, page 68   (Pattern Blocks)
4. Birthday Cakes, page 73   (Circular Geoboards and Geoboards)

The lessons in this cluster explore the use of patterns in solving problems. Each lesson provides a different scenario in which patterns play a role in the solution process. The activities can be worked on in any order, although it is suggested that *Birthday Cakes* be the last activity.

1. Count Square and Countess Triangle *(Using patterns to solve problems)*

In this activity, students develop strategies for solving two problems, one involving squares, and the other, triangles. In searching for solutions, students come to realize that it may be useful to look for patterns that may help solve the problems.

As is true of many activities of this type, the quest for solutions is more important than the solution itself. Teachers should make much of the discussion following each *On Their Own*, allowing for sharing of strategies and observations.

2. Greek Border Designs *(Exploring repetitive patterns)*

In this activity, students analyze design sequences containing repetitive patterns, and use what they learn to create their own such sequences. *On Their Own* Part 1 asks students to use the patterns they find to make predictions about designs containing greater numbers of repetitions. In Part 2, students search for different solutions to a problem involving designs containing repetitive patterns.

3. Table for 63, Please *(Exploring patterns generated by real-world logistics)*

Through their investigation of a problem involving the seating of increasing numbers of people around successively larger tables, students learn how perimeter is affected by the adjoining of congruent shapes.

In the discussion following *On Their Own* Part 1, teachers will want to prompt students to investigate *why* the patterns grow as they do. A full explanation is provided in *Where's the Mathematics?* (page 70).

On Their Own Part 2 provides an open-ended activity in which students use a combination of creativity and reasoning to solve a related problem.

4. Birthday Cakes *(Exploring the role of patterns in inductive and deductive reasoning)*

This activity invites students to explore two similar problems, only one of which generates a predictable pattern. It also provides an informal introduction to inductive and deductive reasoning.

Where's the Mathematics? (page 75) explains the distinctions between the two problems and provides words of caution about drawing conclusions based on a limited number of examples. It also provides a logical line of deductive reasoning that can be used to prove that the pattern generated by the second problem will continue in a predictable way.

COUNT SQUARE AND COUNTESS TRIANGLE

- Spatial visualization
- Pattern recognition
- Square and triangular numbers

Getting Ready

What You'll Need

Color Tiles, 64 per pair

Color Tile grid paper, page 120

Pattern Blocks, 36 green triangles per pair

Pattern Block triangle paper, page 122

Scissors (optional)

Activity master, page 109

Overview

Using Color Tiles, students search to find as many squares as they can on an 8-by-8 checkerboard. They then use green Pattern Block triangles to explore a similar problem involving equilateral triangles. In this activity, students have the opportunity to:

- strengthen visual awareness
- use patterns to solve problems
- investigate the sequence of perfect squares
- investigate sequences that may not be easily discernible

Other *Super Source* activities that explore these and related concepts are:

Greek Border Designs, page 63

Table for 63, Please, page 68

Birthday Cakes, page 73

The Activity

On Their Own (Part 1)

> Count Square and Countess Triangle enjoy challenging each other with brain teasers. It is the Count's turn to pose one of these mind ticklers to the Countess. He chooses the following problem for her to ponder:
> How many squares are there on an 8-by-8 checkerboard?
> Can you help the Countess find the solution?
>
> - Working with a partner, use Color Tiles to build an 8-by-8 square to represent the checkerboard.
> - Decide on a way to search for all the different-sized squares that are on the checkerboard.
> - Record the sizes of the squares you find and the number of squares of each size.
> - Look for patterns that might help in your search.
> - Determine the total number of squares on the checkerboard. Be ready to tell about how you organized your search.

Thinking and Sharing

Invite students to share their solutions and approaches. Then have them help create a class chart listing the dimensions of the squares they found and the number of squares of each size. Discuss and resolve any differences in results.

Use prompts like these to promote class discussion:

- What strategy or strategies did you use to search for squares?
- Were some squares harder to find than others? If so, which ones and why?
- Did you discover any patterns that helped you solve this brain teaser? If so, describe your observations.
- How did you determine that you had accounted for all the squares?
- Do you see any patterns in the class chart? If so, describe them.

On Their Own (Part 2)

What if... the Countess challenges Count Square with a similar problem based on finding the number of triangles in a large equilateral triangle made from Pattern Blocks? Can you help the Count solve his mind tickler?

- Working with a partner, use green Pattern Block triangles to build an equilateral triangle that has a base equal in length to the sides of 6 green triangles.
- Decide on a way to search for all the different-sized triangles contained in your equilateral triangle.
- Record the lengths of the bases of the triangles you find and the number of triangles of each size.
- Look for patterns that might help in your search.
- Determine the total number of triangles in the large triangle. Be ready to tell about the strategies you used to search for triangles and why you think you've found them all.

Thinking and Sharing

Use an approach similar to that used in the first activity to collect data for the class chart. Ask students to tell about their strategies and observations. Although some students may have had difficulty finding and extending a pattern in their chart, encourage them to discuss their attempts and their thoughts.

Use prompts like these to promote class discussion:

- What strategy or strategies did you use to search for triangles?
- Were some triangles harder to find than others? If so, which ones and why?
- What kinds of triangles were you able to find? Why do you think all of the triangles were this same type (that is, equilateral)?

©Cuisenaire Company of America, Inc. COUNT SQUARE, COUNTESS TRIANGLE ◆ Patterns/Functions ◆ Grades 7-8

- How would you describe the orientations of the triangles you found?
- Did you discover any patterns that helped you solve this brain teaser? If so, describe your observations.
- How did you determine that you had accounted for all the triangles?
- Do you see any patterns in the chart? If so, describe them.
- Whose brain teaser do you think was more challenging, the Count's or the Countess's? Why?

For Their Portfolio

Write a letter to the Count and Countess explaining how you would go about finding all the squares on a 20-by-20 Color Tile checkerboard.

Teacher Talk

Where's the Mathematics?

To find all the squares on a checkerboard, students must realize that they need to look beyond the obvious. They must recognize that in addition to the 64 small squares, there are many other squares, ranging in size from 2-by-2 squares to an 8-by-8 square. This recognition may help them see the need for organizing their search.

Students may go about their search in different ways. Some students may choose to count random squares of a particular size, but may quickly discover that it is difficult to know if they've accounted for every square. Other students may investigate the problem by looking at the total number of squares contained in smaller Color Tile squares, such as 1-by-1 squares, 2-by-2 squares, 3-by-3 squares, and so on. With each larger square they build, students can look for patterns that may help them determine the total number of squares contained in any size Color Tile square.

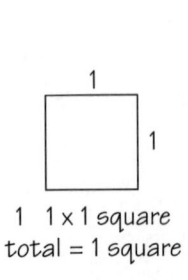

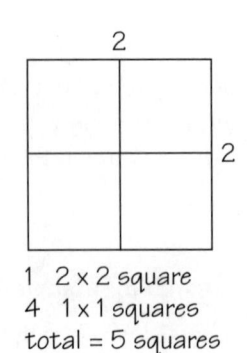

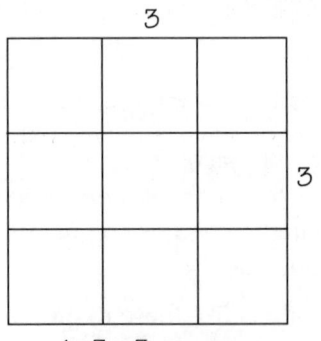

1 1 x 1 square
total = 1 square

1 2 x 2 square
4 1 x 1 squares
total = 5 squares

1 3 x 3 square
4 2 x 2 squares
9 1 x 1 squares
total = 14 squares

Another approach students might take is to cut out squares of different sizes (1-by-1, 2-by-2, 3-by-3, etc.) from Color Tile grid paper and use these squares to help track and count squares of different sizes. This method is particularly useful for students who have trouble seeing the overlapping squares. Placing one cutout square at a time in the upper left-hand corner of the checkerboard, students can slide the cutout, tile-by-tile, through the square to determine the number of each particular-sized square. The first few steps of this method are shown below.

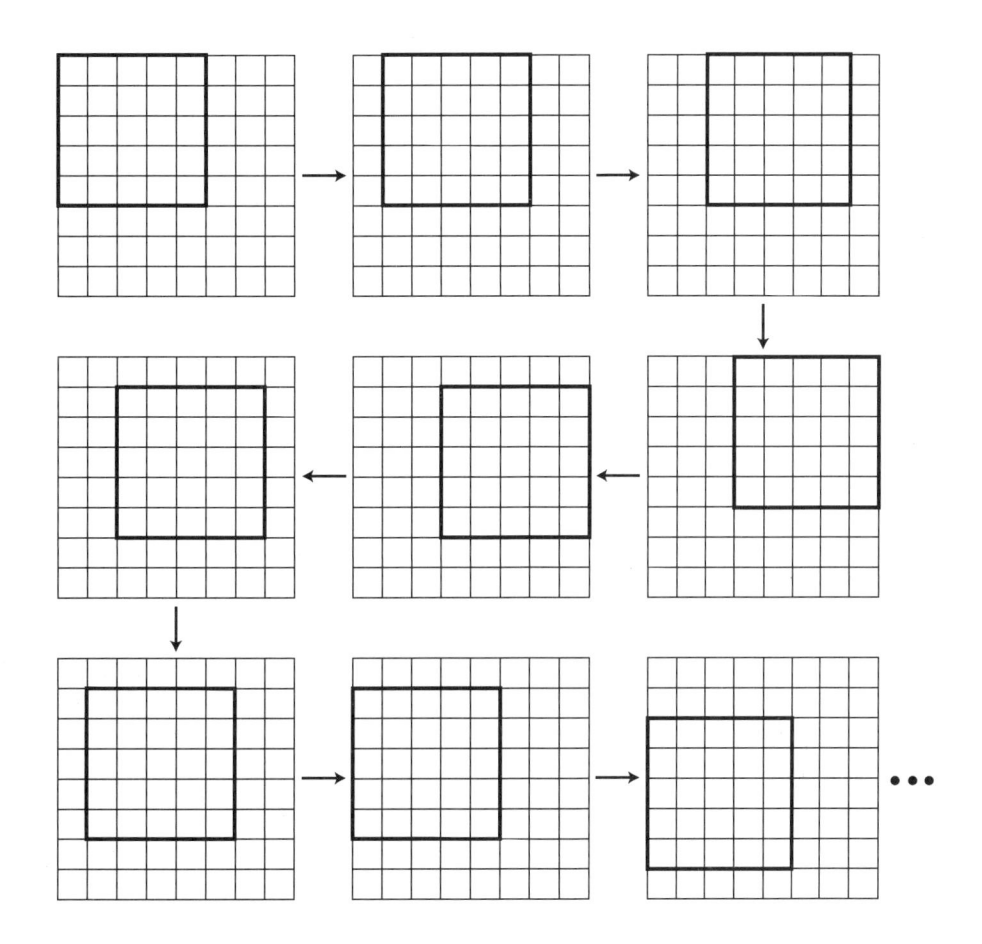

As students organize their data, they can look for patterns that may help them predict how many squares of different sizes they can expect to find. If they have begun by counting larger squares first, they may have created a chart like the one shown at the right.

After collecting and organizing the data for the first few different-sized squares, students may notice that the number of squares of each size is always a perfect square. Once the progression of consecutive perfect squares becomes evident, students may hypothesize about the numbers of successively smaller squares, confirm their hypotheses by counting, and use the sequence of perfect squares to complete their charts.

size of square	number of squares
8 x 8	1
7 x 7	4
6 x 6	9
5 x 5	16
4 x 4	25
3 x 3	36
2 x 2	49
1 x 1	64

Students who are not familiar with the sequence of perfect squares may need to continue searching for the number of squares until they see that the numbers in the second column of the chart increase by consecutive odd integers (3, 5, 7, ...). Once this pattern is established, students can use it to predict how many squares of each size they should expect to find on the checkerboard.

Students may also point out another pattern: The product of the dimensions of the largest square (an 8-by-8) is the number of squares of the smallest size that can be found (64); the product of the dimensions of the second largest square (a 7-by-7) is the number of squares of the second smallest size that can be found (49), and so on. The sum of these products, 204, is the total number of squares on the checkerboard.

To solve the triangle problem, students may try many of the same techniques they used to solve the square problem. Students may find this problem more challenging than the first because not only do they have to look for overlapping triangles, but they must also look for triangles in two different orientations: one where a vertex of the triangle points upward and one where a vertex of the triangle points downward.

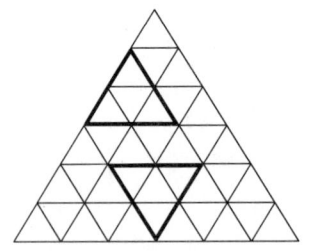

The data for this part of the activity is shown below. Students may not notice any patterns in the data while they're working, and may need to systematically count triangles each step of the way to obtain the total number of triangles (78). Other students may have a hunch that since the problem involving squares led to a pattern of square numbers, this problem, involving triangles, may lead to a pattern involving triangular numbers. Indeed, the first three numbers in the second column (1, 3, 6) would seem to confirm this hunch; however, the pattern falls apart with the number of equilateral triangles with side length 3 units.

side length of triangle	number of triangles	
6	1	
5	3	
4	6	
3	11	← 10 △'s + 1 ▽
2	21	← 15 △'s + 6 ▽'s
1	36	← 21 △'s + 15 ▽'s

Some students may notice that when the data for triangles with sides of 3, 2, and 1 units are broken down by triangle orientation, an observable pattern emerges. The numbers of triangles oriented the same way as the large triangle continue the triangular number sequence, while the numbers of triangles oriented in the opposite position form the sequence of every other triangular number. Although students may not have been able to identify and use this pattern during their search (which may have lead them to feel that the Countess's problem was more challenging than the Count's), it is important for them to see that there is indeed an identifiable and explainable pattern, as their intuition may have led them to believe.

GREEK BORDER DESIGNS

- Visual sequences
- Using patterns to make predictions
- Writing algebraic expressions
- Interpreting data

Getting Ready

What You'll Need

Color Tiles, at least 30 yellow, 40 red, and 45 blue per pair

Colored pencils or crayons (yellow, red, and blue)

Color Tile grid paper, page 120

Activity master, page 110

Overview

Students determine how to calculate the number of Color Tiles needed for designs in a sequence without actually building each design. They then create their own sequence of designs based on given patterns. In this activity, students have the opportunity to:

- look for patterns in numbers and designs
- collect and analyze data
- use patterns to make predictions
- write algebraic expressions to generalize patterns

Other *Super Source* activities that explore these and related concepts are:

Count Square and Countess Triangle, page 58

Table for 63, Please, page 68

Birthday Cakes, page 73

The Activity

On Their Own (Part 1)

Almost 2,500 years ago, Greek artisans decorated their pottery with a geometric border called the Greek Key design. Can you predict how many tiles of each color will be needed to build any design in a sequence of Greek Key designs?

- Look at this sequence of Greek Key designs built using red, yellow, and blue Color Tiles.

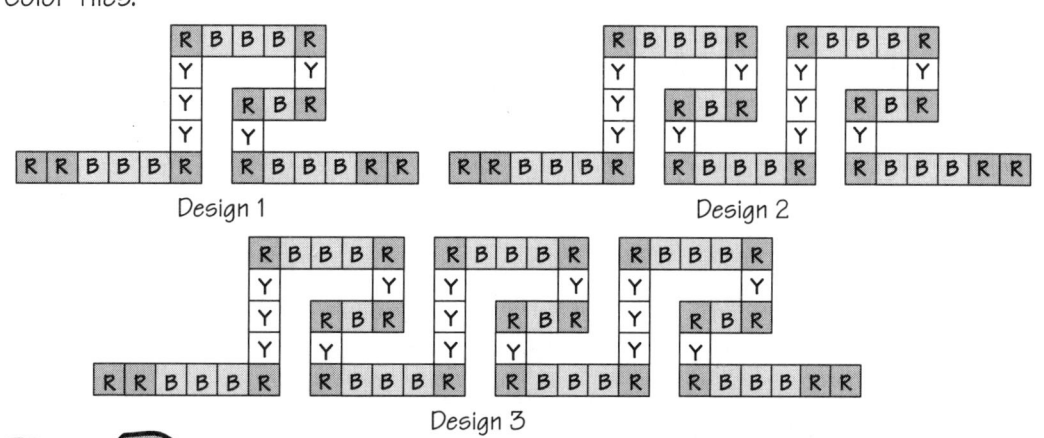

Design 1

Design 2

Design 3

- Working with a partner, record the number of red, yellow, and blue tiles used in each design. Also record the total number of tiles in each design. Organize your data in a chart.
- Use your Color Tiles to build the next few terms of the sequence and add the information about these terms to your chart. Continue until you see patterns in your data.
- Predict how many tiles of each color will be needed for the 10th design in the sequence (without building it). Do the same for the 100th design.
- Predict how many tiles of each color will be needed for the *n*th design in the sequence. Express your predictions using algebraic expressions.
- Be ready to talk about how you made your predictions.

Thinking and Sharing

Invite students to share their predictions and describe their reasoning. Have them help you complete a class chart with these column headings: *design number, number of red tiles, number of yellow tiles, number of blue tiles,* and *total number of tiles.* Include data for at least six designs in the sequence.

Use prompts like these to promote class discussion:

- How would you describe the pattern(s) formed by the sequence of Greek Key designs?
- How did you go about building successive terms in the sequence?
- What patterns did you find in your data?
- How did you predict the number of red, yellow, and blue tiles needed for the 10th design? the 100th design? the *n*th design?
- What algebraic expressions did you write to generalize your findings?

On Their Own (Part 2)

What if... *the number of tiles of each color is limited, and the sequence of designs must contain particular patterns of tile colors? What Greek border designs can be created?*

- Study the chart at right.
- Working with your partner, create a sequence of designs that fits the data in the chart. To do this, first determine the number of red, yellow, and blue tiles that are added to each design to produce the next design in the sequence. These tiles make up what is called the "repeat" portion of the design. As its name suggests, the repeat will appear over and over again in your designs.

design number	number of red tiles	number of yellow tiles	number of blue tiles	total number of tiles
1	6	4	3	13
2	7	8	6	21
3	8	12	9	29
4	9	16	12	37
5	10	20	15	45
6	11	24	18	53

- Create a design using the tiles in the repeat. Add on any additional tiles needed to satisfy the data for design 1.

- Make two copies of your repeat and link them together end to end. Add on any additional tiles indicated by the data for design 2. Try to maintain pattern and color symmetry at both ends of your border design.

- Build three more terms in your sequence. Record and color your design on Color Tile grid paper.

- Create several other Greek border designs using the same data. Record your designs on grid paper and compare them.

Thinking and Sharing

Have each pair of students post one or two of their Greek border designs on the board. Ask students to verify that the designs fit the data. (Note: You may prefer to have pairs reconstruct one of their design sequences with Color Tiles and have students walk around the room to look at and verify each other's designs.)

Use prompts like these to promote class discussion:

- What was hard about making your Greek border designs? What was easy?
- How did you use the data to build your sequence?
- How did you decide on a design for the repeat?
- How many different border designs did you create? How are they alike? How are they different?
- How are the designs built by other pairs different from the ones you built?

For Their Portfolio

Create you own Greek border design using Color Tiles. Record and color it on Color Tile grid paper. Then write a paragraph or two describing the underlying patterns in your design. Tell how the numbers of tiles needed for longer versions of your border design can be calculated.

Teacher Talk

Where's the Mathematics?

Students may have different ways of describing what they see when they look at the Greek Key design. Encourage them to share their observations. Data for the first 10 designs of the sequence are shown in the chart on the following page.

design number	number of red tiles	number of yellow tiles	number of blue tiles	total number of tiles
1	10	5	10	25
2	16	10	17	43
3	22	15	24	61
4	28	20	31	79
5	34	25	38	97
6	40	30	45	115
7	46	35	52	133
8	52	40	59	151
9	58	45	66	169
10	64	50	73	187
⋮	⋮	⋮	⋮	⋮
n	$6n + 4$	$5n$	$7n + 3$	$18n + 7$

In studying the data within the columns, students can observe that the entries in the *number of red tiles* column increase by 6, those in the *number of yellow tiles* column increase by 5, those in the *number of blue tiles* column increase by 7, and those in the *total number of tiles* column increase by 18. By continuing the patterns found within the columns, students can determine the data for the tenth design in the sequence. They may agree, however, that it would be tedious to use this method to find data for higher numbered terms in the sequence, and that using algebraic expressions would be much more efficient and desirable.

Some students may recognize that the entries in the *number of yellow tiles* column are multiples of 5 and can be expressed as 5 times the design number. Thus, the number of yellow tiles in the *n*th design can be represented using the expression $5n$.

To determine the number of red tiles based on the design number, students may experiment with multiples of 6 (since the entries differ by 6) and compare the results to the values in the *number of red tiles* column. For example, comparing 6, 12, 18, 24, 30, 36, 42, 48, 54, and 60 to the ten entries in the column, students may discover that each multiple is 4 less than the actual number of red tiles. By adding 4 to the appropriate multiple of 6, the number of red tiles can be found.

design number		number of red tiles
1	1(6) + 4 =	10
2	2(6) + 4 =	16
3	3(6) + 4 =	22
⋮	⋮	⋮
10	10(6) + 4 =	64
⋮	⋮	⋮
n	$n(6) + 4 =$	$6n + 4$

In the same manner, to determine the number of blue tiles, students may compare multiples of 7 (7, 14, 21, 28, 35, 42, 49, 56, 63, and 70) to the ten entries in the *number of blue tiles* column. In doing so they should find that each multiple is 3 less than the actual number of blue tiles. Thus, by adding 3 to the appropriate multiple of 7, the number of blue tiles can be found.

design number		number of blue tiles
1	1(7) + 3 =	10
2	2(7) + 3 =	17
3	3(7) + 3 =	24
⋮	⋮	⋮
10	10(7) + 3 =	73
⋮	⋮	⋮
n	$n(7) + 3 =$	$7n + 3$

Several methods can be used to find the expression that generates the values in the *total number of tiles* column. Some students may use the same method described above. In comparing the entries to multiples of 18 (since the entries differ by 18), students should find that each corresponding multiple is 7 less than the actual total. They can then conclude that the algebraic expression $18n + 7$ will generate the total number of tiles for the *n*th design.

Another method for finding the total number of tiles in the *n*th design involves adding together the algebraic expressions representing the number of red, yellow, and blue tiles needed for the *n*th design: $(6n + 4) + (5n) + (7n + 3) = 18n + 7$.

In the second activity, students have the opportunity to analyze data presented in a chart and build geometric models representing that information. Numerous border designs can be generated using the given data, and students should be encouraged to use their imagination and artistic creativity to come up with as many variations as possible.

Students should find that the repeat consists of 1 red tile, 4 yellow tiles, and 3 blue tiles. Students may create a variety of repeat designs; some may have a random arrangement of Color Tiles, while others may have different types of color and/or pattern symmetry.

In building the first design of the sequence, students will need to use 6 red tiles, 4 yellow tiles, and 3 blue tiles. When these numbers are compared to the numbers of each color of tile used in the repeat, students should realize that this first design contains 5 red tiles that are not part of the repeat. These 5 tiles will need to be placed at the outside ends of each design. The placement of these tiles may become clearer when students build the second design. After linking together two copies of the repeat, these 5 red tiles must again be added on. In order to maintain pattern and color symmetry for all terms of the design sequence, students must come to realize that a red tile in the repeat must be located at one end of the repeat. Shown below are designs 1 and 2 of several possible design sequences. The repeat of each sequence is highlighted.

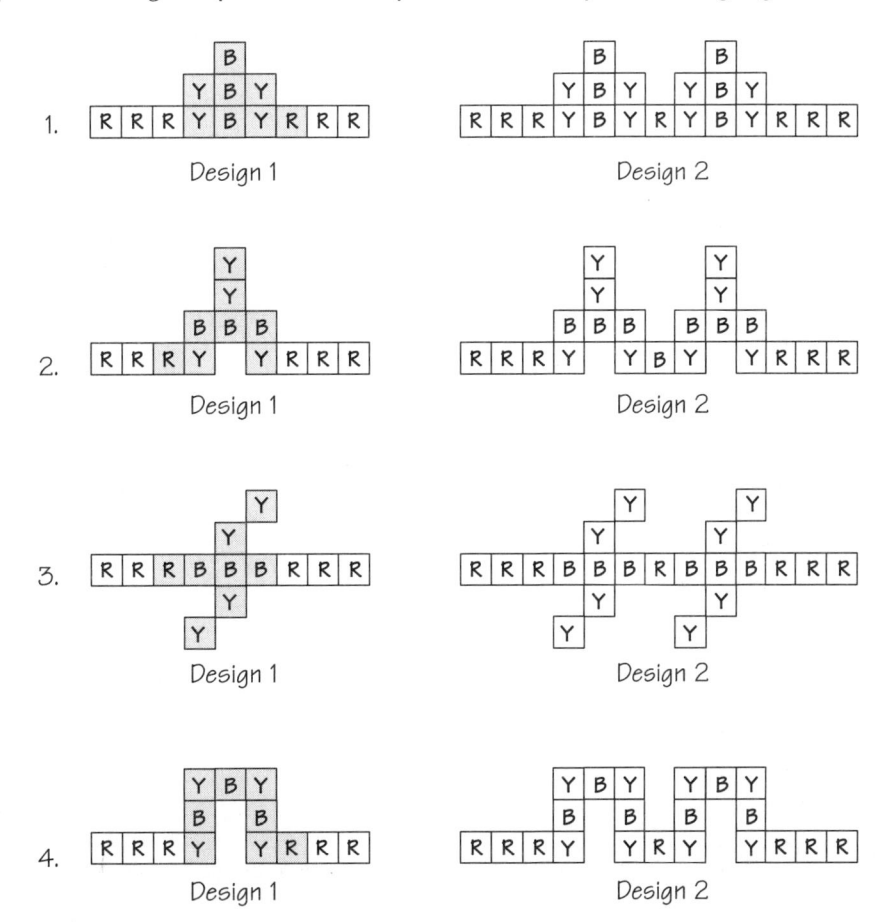

© Cuisenaire Company of America, Inc. GREEK BORDER DESIGNS ◆ Patterns/Functions ◆ Grades 7-8

TABLE FOR 63, PLEASE

- Perimeter
- Organizing and interpreting data
- Pattern recognition
- Writing algebraic expressions

Getting Ready

What You'll Need

Pattern Blocks, 1 set per pair

Activity master, page 111

Overview

Students use Pattern Blocks to investigate how perimeter changes as blocks are added to a shape. In this activity, students have the opportunity to:

- collect and organize data
- recognize and use patterns to make predictions
- write algebraic expressions to generalize patterns

Other *Super Source* activities that explore these and related concepts are:

Count Square and Countess Triangle, page 58

Greek Border Designs, page 63

Birthday Cakes, page 73

The Activity

On Their Own (Part 1)

> Matt and Jamal own a company that rents tables and chairs for parties. They have 100 tables of each shape in their inventory. These tables can be joined together to provide seating accommodations for larger groups of people. Can you predict how many people can sit at one large table made by joining 100 tables?
>
> - Work with a partner. Use a green Pattern Block triangle to represent one table. Figure that one person can be seated on each side of a triangular table.
>
> - Build larger tables by joining triangles side to side. Add tables in such a way so that each new table added shares only one side with the existing table. Record how many people can sit at each larger table. Make a chart to keep track of your findings.
>
> - Continue until you have built at least 6 or 7 tables of different sizes.
>
> - Look for patterns in the numbers you recorded. Use these patterns to predict how many people can sit at a table built from 100 triangular tables. Then generalize your findings and predict how many can sit at *n* triangular tables.

- Repeat the process of building larger tables, first using orange Pattern Block squares as tables, and then using yellow Pattern Block hexagons. Be sure that each new shape added shares only one side with the existing table.
- For each shape, look for patterns that would help you predict how many people can sit at a table built from 100 such tables, and then generalize your findings to predict how many can sit at *n* tables.

Thinking and Sharing

Invite students to share their charts for the different-shaped tables and discuss their findings.

Use prompts like these to promote class discussion:

- How did you go about building the tables?
- What patterns did you discover in the numbers you recorded?
- Why do you think each pattern grows in the way that it does?
- How many people can sit at the table built from 100 triangular tables? from 100 square tables? from 100 hexagonal tables? Explain how you know.
- How many people cat sit at the table built from *n* triangular tables? from *n* square tables? from *n* hexagonal tables? Explain how you know.

On Their Own (Part 1)

What if... Matt and Jamal are asked to provide tables and chairs for a party Rosie is having for 63 people? Using any combination of triangular, square, and hexagonal tables, how can they provide seating for exactly 63 people?

- Consider seating patterns that might be generated by combining a variety of different-shaped tables. Use the materials and information you gathered from the first activity.
- Look for table combinations that can accommodate exactly 63 people. Build or sketch models of your arrangements.
- Decide whether any of the table combinations would be more desirable than others. Be ready to explain your reasoning.

Thinking and Sharing

Have students display their models and/or post their sketches. Ask them to tell about the thinking they used in designing their table arrangements.

Use prompts like these to promote class discussion:

- How did you go about modeling or sketching seating arrangements for 63 people?
- What was hard about finding table combinations for a specific number of people?
- How did you use what you had discovered in the first activity to help in solving this problem?
- What table combinations did you find that would accommodate 63 people?
- What combination of tables did you determine to be the most desirable? Why?

For Their Portfolio

Write a letter to Matt and Jamal describing the combination of triangular, square, and hexagonal tables you think they should supply for Rosie's party of 63 people and why you think it is the best choice. Include any diagrams or instructions that might be helpful.

Teacher Talk

Where's the Mathematics?

As students build larger tables and count the number of seats they will accommodate, they begin to see a relationship between the geometric growth of a shape (the number of tables), and the growth of its perimeter (the number of seats). Although students are working with triangular, square, and hexagonal tables, they may be surprised to see that the perimeters do not increase by increments of 3, 4, and 6, respectively. Some students, realizing that one seat from the original table arrangement is being forfeited as the new table is attached, may think that the seat or perimeter increments may be 2, 3, and 5. This is not the case either.

The change in seating capacity can best be seen and understood by building larger tables and observing what occurs as each new table is added, as shown below.

For each triangular table added to the existing larger table, 1 seat of the existing table is forfeited and 2 new seats are added. The net change is 1 additional seat.

For each square table added to the existing larger table, 1 seat of the existing table is forfeited and 3 new seats are added. The net change is 2 additional seats.

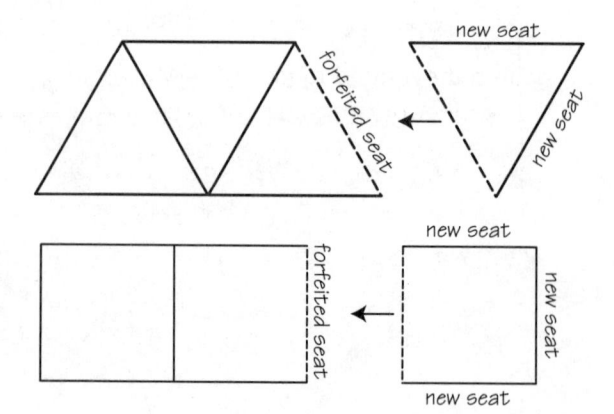

For each hexagonal table added to the existing larger table, 1 seat of the existing table is forfeited and 5 new seats are added. The net change is 4 additional seats.

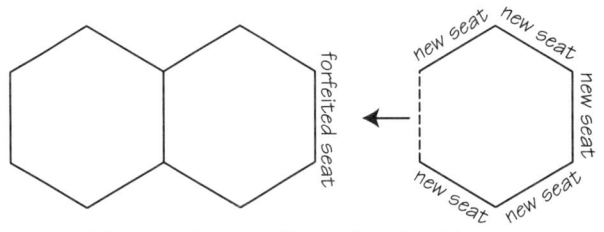

To determine the number of people who can be seated at 100 tables, students will need to build several examples of table arrangements for each Pattern Block shape. They may organize their data in charts like those shown here:

number of triangular tables	number of seats
1	3
2	4
3	5
4	6
5	7
6	8
7	9
⋮	⋮
100	102
⋮	⋮
n	$n + 2$

number of square tables	number of seats
1	4
2	6
3	8
4	10
5	12
6	14
7	16
⋮	⋮
100	202
⋮	⋮
n	$2n + 2$

number of hexagonal tables	number of seats
1	6
2	10
3	14
4	18
5	22
6	26
7	30
⋮	⋮
100	402
⋮	⋮
n	$4n + 2$

Some students may extend the sequences they see in their charts, adding the common difference to successive terms to find the 100th term. Other students may be able to make a connection between the number of tables and the number of seats. For example, students may notice that the number of people who can sit at a table made from a certain number of triangular tables is 2 more than the number of tables. From there, they can determine that the number of people that can be seated at a table made from 100 triangular tables is 100 + 2, or 102. In general, n triangular tables will form a table large enough to accommodate $n + 2$ people.

Similarly, students may notice that the number of people who can sit at a table made from a certain number of square tables is 2 more than twice the number of tables. From there, they can determine that the number of people that can be seated at a table made from 100 square tables is 2(100) + 2, or 202. In general, n square tables will form a table large enough to accommodate $2n + 2$ people.

The relationship for hexagonal tables may be a bit more difficult for students to find; however, the pattern they found for square tables may provide some ideas for them to try. They should find that the number of people who can sit at a table made from a certain number of hexagonal tables is 2 more than four times the number of tables. From there, they can determine that the number of people that can be seated at a table made from 100 hexagonal tables is 4(100) + 2, or 402. In general, n hexagonal tables will form a table large enough to accommodate $4n + 2$ people.

The second activity allows students to use mathematical reasoning and artistic creativity to explore table combinations that will accommodate 63 people. Students may realize that 61 triangular tables will accommodate 63 seats; however, they may argue that supplying 61 tables may not be efficient for Matt and Jamal. Thus, students should explore other combinations of different-shaped tables that will provide seating for 63 people.

©Cuisenaire Company of America, Inc.   TABLE FOR 63, PLEASE ◆ Patterns/Functions ◆ Grades 7-8

Some students may use their charts to find a large table made from tables of one shape that comes close to seating 63 people. For example, 15 hexagonal tables seat 62 people. They may then realize that by adding one triangular table, they can accommodate the 1 extra seat. This accomplishes the job using the minimum number of tables, 16. This combination may be more efficient for Matt and Jamal, although the hexagonal tables may be heavier to move than the others. The resulting arrangement of tables might also be less conducive to a party setting.

Some students may decide that using only one table of a different shape from all the others (as described above) may not be aesthetically pleasing. They may look for a combination that uses several tables of each shape, as in the example below.

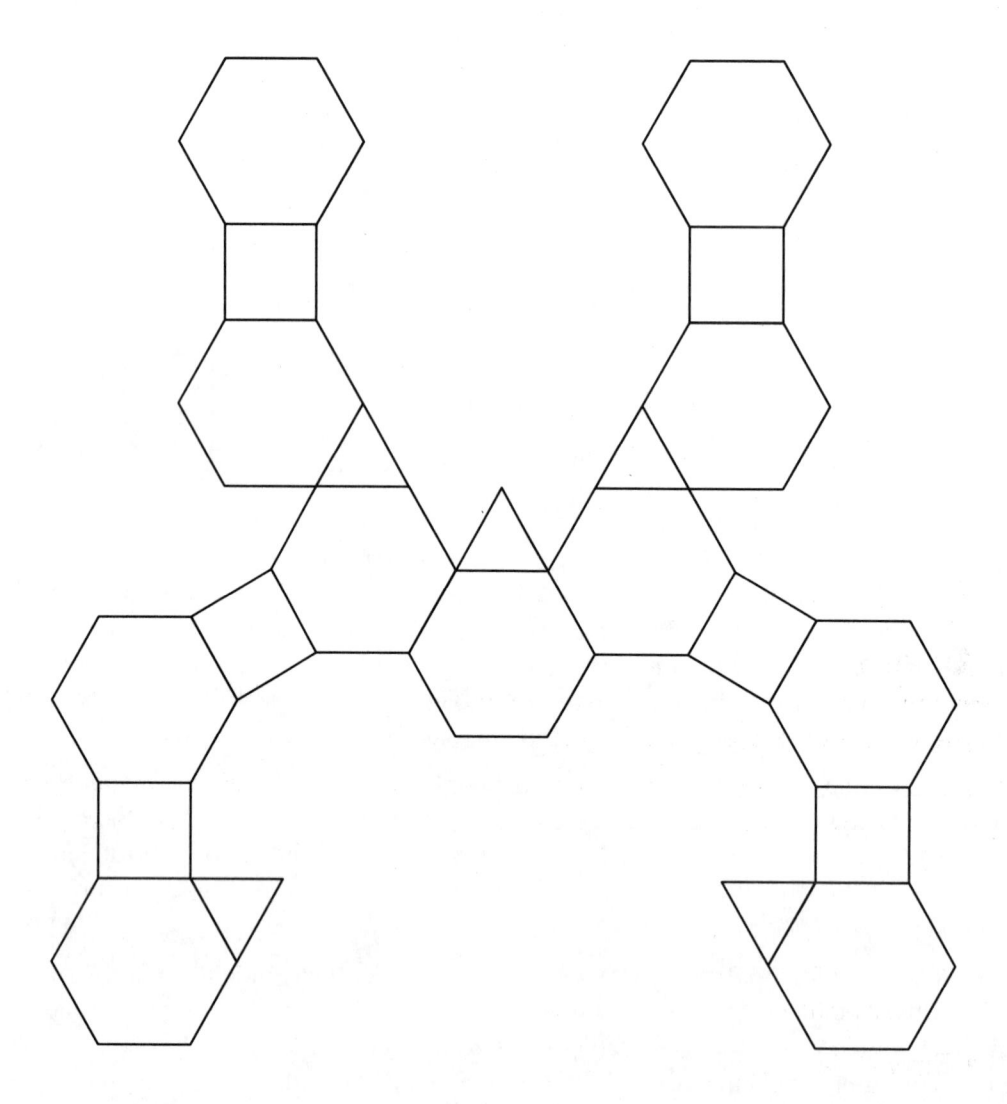

BIRTHDAY CAKES

- Collecting and organizing data
- Pattern recognition
- Inductive and deductive reasoning
- Writing algebraic expressions

Getting Ready

What You'll Need

Circular Geoboard, 1 per student

Geoboard, 1 per student

Rubber bands

Circular geodot paper, page 118

Geodot paper, page 117

Dot paper, page 119

Activity master, page 112

Overview

Using a circular Geoboard, students look for patterns in the number of regions formed when points on the circumference of a circle are connected. They then extend their investigation to Geoboard rectangles. In this activity, students have the opportunity to:

- collect and organize data
- make predictions and verify hypotheses
- write algebraic expressions to generalize patterns
- learn about the limitations of inductive reasoning

Other *Super Source* activities that explore these and related concepts are:

Count Square and Countess Triangle, page 58

Greek Border Designs, page 63

Table for 63, Please, page 68

The Activity

On Their Own (Part 1)

Anastasia is having a birthday party and is preparing to slice her round birthday cake. She wonders if there is a way to predict the number of pieces of cake that can be obtained by cutting the cake a certain way. Can you help Anastasia determine how many people she can serve?

- Using the Circular Geoboard to represent Anastasia's birthday cake, join any two points on the circle with a rubber band representing the first cut. (Do not use the center point on the Geoboard as one of the two points.)

- Draw a diagram of the cake and the first cut on circular geodot paper. Record the number of points connected, the total number of cuts made (in this case, 1), and the number of servings of cake obtained.

- Choose a third point on your Geoboard cake and make a series of cuts using rubber bands that join your new point to the points you used before. Select this new point in a way that will maximize the number of regions formed, producing as many servings as possible.

- Record your cake and data as you did before.

- Repeat the above procedure, adding a 4th point and then a 5th point. Be sure to record a diagram and the data for each point added.

- Organize your data and look for patterns that can help you predict the maximum number of servings of cake that will be formed by adding a 6th point. Then test your prediction by adding a 6th point, connecting it to all the others, and counting the number of pieces of cake obtained.

- Share and compare your work with your partner. Be ready to discuss your findings.

Thinking and Sharing

Have students help you organize the data for a class chart. Have them share their observations, telling about their predictions and their findings.

Use prompts like these to promote class discussion:

- How did you decide which point to add each time?
- How did moving the position of the newly chosen point affect the number of cuts? the number of pieces of cake?
- What patterns did you find in the data?
- What predictions did you make about the maximum number of pieces of cake that could be formed by adding a 6th point?
- How did your actual results compare with your prediction? If they differed, how might you explain the discrepancy?

On Their Own (Part 2)

What if... Anastasia thinks that at her next birthday party it would be easier for her to serve a rectangular cake? Would she find that the relationship between the number of cuts and the number of slices is different from what she found with her circular cake?

- On your Geoboard, use a rubber band to mark off a rectangle whose length is 1 unit longer than its width. Let this rectangle represent Anastasia's birthday cake. Use rubber bands to "slice" the cake into 1-by-1 squares.

- Copy your cake onto geodot paper. Record its dimensions, the number of cuts made, and the total number of servings of cake obtained.

- Make several different Geoboard cakes, each with length 1 unit longer than the width. Cut each cake into single square servings as described. Make a drawing of your cake and record the data as you did with your first cake.

- Organize your data in a chart and look for patterns.

- Generalize your findings by writing algebraic expressions that can be used to predict the number of cuts and the number of pieces of cake for rectangular cakes with width *n* units and length 1 unit longer. Use dot paper to test your hypotheses on other rectangular cakes whose lengths are 1 unit longer than their widths. Be ready to discuss your findings.

Thinking and Sharing

Invite students to share their findings. Have them help you organize their data in a class chart with headings *dimensions of cake, number of cuts,* and *number of pieces of cake.* Ask them also to share the examples they used to check their hypotheses and add the data from these examples to the chart.

Use prompts like these to promote class discussion:

- What patterns did you notice in your data?
- What algebraic expressions did you write to represent the patterns?
- What happened when you tested your hypotheses on other rectangles?
- Do you think your expressions will work for all rectangular cakes whose lengths are 1 unit longer than their widths? Why or why not?
- How can you explain the differences between the first and second parts of this activity?

For Their Portfolio

Describe how the two parts of this activity were alike and how they were different. Discuss the roles of patterns and reasoning. Include any conclusions you can make about using patterns to make predictions.

Teacher Talk

Where's the Mathematics?

The two parts of the activity allow students to search for patterns, make predictions, and test hypotheses. By considering two similar types of problems and their outcomes, students may begin to understand why they must use caution when drawing generalized conclusions from a limited set of observations.

As they start to compile their data in Part 1, students may recognize the sequence formed by the entries in the *number of cuts* column as the triangular number sequence. (For more on this sequence, see *Backyard Improvements,* page 9.) No matter how they choose their points, students should find that the number of cuts for a given number of points is always the same. They should also find that for fewer than six points, the number of pieces of cake formed by a certain number of cuts is always the same, no matter which points are chosen.

number of points connected	number of cuts	number of pieces of cake
2	1	2
3	3	4
4	6	8
5	10	16
6	15	30 or 31

After connecting four points, counting the regions obtained, and observing the doubling pattern in the *number of pieces of cake* data, students may be able to predict correctly that the maximum number of regions formed by connecting five points will be 16. However, when they try joining six points in every possible way, they will find either 30 regions (if three diagonals intersect at the center) or 31 regions (the maximum number possible). No matter how students select the six points on the circular Geoboard, they will not be able to produce the 32 regions they may have expected based on what appears to be a doubling pattern. The pattern has fallen apart!

Some students may think they've made a counting error, or that there must be a way to produce the 32 regions but they haven't picked the right combination of points. These students may want to investigate the next case, joining seven points, and checking to see if this produces the expected 64 regions. If they do this, they should find that the maximum number of regions is 57, providing further verification of the disintegration of what appeared to be a doubling pattern.

This activity provides an opportunity for students to see that inductive reasoning based on a limited number of examples is not always a reliable basis for drawing conclusions. Students should understand that sometimes it works, sometimes it does not. Deductive reasoning, on the other hand, draws conclusions based on statements that have been accepted as true. Deductive reasoning is based on facts and logic and is always valid if the original assumptions are valid.

When students approach the second part of the activity, they may suspect that it, too, will produce an initial pattern that will disintegrate after several terms. As they start to gather information about the dimensions of the rectangular cake, the number of cuts, and the number of 1-by-1 square pieces formed, they will begin to see patterns emerging that are valid regardless of how many examples are considered.

Students may notice that the entries in the *number of cuts* column form the sequence of odd numbers. They may also figure out that the entries in the last column increase by consecutive even integers. If students think about *why* these patterns exist (in terms of what's happening geometrically), they can be confident that these patterns will continue in a predictable way.

Using logic and basic algebraic facts, students can use the following line of deductive reasoning to confirm their generalizations:

dimensions of cake	number of cuts	number of pieces of cake
1 × 2	1	2
2 × 3	3	6
3 × 4	5	12
4 × 5	7	20
5 × 6	9	30
(add data from students' examples)		
$(n) \times (n+1)$	$2n - 1$	$n(n+1)$, or $n^2 + n$

> The dimensions of any rectangle whose length is 1 more than its width can be represented by n and $(n + 1)$. If the width is n units, then the minimum number of cuts needed to separate them is $(n - 1)$, and if the length is $(n + 1)$ units, then the minimum number of cuts needed to separate them is n. Therefore, the total number of cuts needed to separate the square pieces of cake is $(n - 1) + n$, or $2n - 1$.
>
> By multiplying the length and width of the cake, the number of square pieces of cake can be found. (Students may realize that this is the same as finding the area of the rectangle.) For a rectangle with dimensions n units by $(n + 1)$ units, the number of unit squares (pieces of cake) is $n(n + 1)$, or $n^2 + n$.

For many students, this activity may be their first taste of the distinction between inductive and deductive reasoning. Make sure that students understand that patterns *are* useful tools for solving problems, but that it's important to keep in mind that a pattern that appears to be predictable may not always be so. Tell them to trust their instincts, but also to look for ways of confirming *why* a particular pattern that appears to be evident should continue in a predictable way.

Investigating Other Topics Using Patterns and Functions

1. Visual Effects, page 78   (Cuisenaire Rods and Tangrams)
2. Pascal Pastimes, page 83   (Geoboards and Color Tiles)
3. Beehive Buzz, page 88   (Pattern Blocks)
4. Carol's Kite Kits, page 93   (Snap™ Cubes)

The lessons in this cluster provide opportunities for students to work on activities that use patterns and functions in different ways. Students will be asked to describe and create sequences, make predictions, use patterns to solve problems, and investigate sequences generated by geometric relationships. The activities can be worked on in any order.

1. Visual Effects (Exploring design sequences)

This activity provides students with an opportunity to strengthen their mathematical vocabulary and communication skills as they create, analyze, and describe design sequences.

Activities of this type motivate the need for precise mathematical terminology. Teachers may want to follow the lesson with an activity in which students work together to generate a list of mathematical terms and their meanings.

2. Pascal Pastimes (Exploring the patterns in Pascal's Triangle)

In this activity, students investigate solutions to a problem involving paths, leading to the discovery of Pascal's Triangle. They then extend their investigation by performing a probability experiment in which the distribution of outcomes produces the patterns found in Pascal's Triangle.

Teachers should encourage students to search for and describe the many patterns that appear in the data. *For Their Portfolio* suggests investigating other applications of Pascal's Triangle.

If Color Tiles are not available, coins (or any other objects that can be flipped fairly) may be used.

3. Beehive Buzz (Exploring patterns formed by equivalence)

In this activity, students develop strategies for solving problems involving equivalence relations. During the solution process, various patterns emerge, enabling students to predict and generate additional solutions.

Teachers may want to extend the activity by using the data to illustrate the connection to equivalence among fractions and among sums of fractions. Students who have worked with Pattern Blocks in a fraction setting are familiar with the use of trapezoids, blue rhombuses, and triangles as representations of (respectively) halves, thirds, and sixths of the yellow hexagon. Thus, a solution that shows 7 hexagons covered by, for example, 10 trapezoids and 12 triangles, is equivalent to a solution using 15 blue rhombuses and 12 triangles.

$$7 = \frac{10}{2} + \frac{12}{6} = \frac{15}{3} + \frac{12}{6}$$

$$\frac{10}{2} = \frac{15}{3}$$

4. Carol's Kite Kits (Reinforcing number sequences)

This activity provides students with an opportunity to reinforce what they've learned about number sequences as they determine the amounts of materials needed to build successively larger kites. Two different types of kites are dealt with, leading to work with different types of patterns and sequences.

The *On Their Owns* assume that students are familiar with ways of finding areas of rectangles and triangles. Students are also asked to generalize their findings by writing algebraic expressions. If this is beyond the students' ability, teachers may elect to have students describe their patterns verbally.

VISUAL EFFECTS

- Spatial visualization
- Pattern recognition
- Mathematical vocabulary
- Writing and interpreting descriptions

Getting Ready

What You'll Need

Cuisenaire Rods, 1 set per pair

1-centimeter grid paper, page 121

Colored pencils or crayons

Tangrams, 2 sets per pair

Activity master, page 113

Overview

Students use Cuisenaire Rods and/or Tangram pieces to build successive terms of given sequences. They then create an original sequence, write rules describing it, exchange rules with another group, and try to build each other's sequences. In this activity, students have the opportunity to:

- use logic and reasoning to extend patterns
- create sequences
- describe sequences in writing
- strengthen mathematical communication skills

Other *Super Source* activities that explore these and related concepts are:

Pascal Pastimes, page 83

Beehive Buzz, page 88

Carol's Kite Kits, page 93

The Activity

On Their Own (Part 1)

Although mathematical and artistic abilities are located in different hemispheres of the brain, you can strengthen both of these abilities by working with visual design sequences. Can you extend the patterns for the two sequences below?

- Study the first 4 terms of Sequence #1.
- Working with a partner, use Cuisenaire Rods to build the next six terms of Sequence #1.
- Draw and color the first ten terms of Sequence #1 on 1-centimeter grid paper.
- A set of rules for building Sequence #2 is given below. Using Cuisenaire Rods and Tangram pieces, work with your partner to build the first eight terms of Sequence #2.

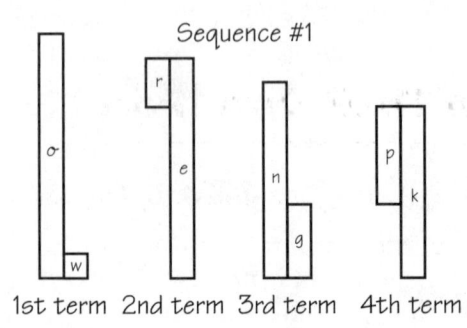

Sequence #1

1st term   2nd term   3rd term   4th term

Sequence #2

Every term contains either a small or a medium Tangram triangle. The small triangle appears in every other term beginning with the first. It sits on one of its shorter sides with its right angle in the lower right corner. The medium triangle, whenever it appears, balances on the vertex at its largest angle. To the right of each triangle is a Cuisenaire Rod positioned horizontally. The rods increase in length by 1 unit from term to term, beginning with a rod whose length is 1 unit. For odd-numbered terms, the rod is attached to the side of the triangle with bottom edges of both blocks in alignment. For even-numbered terms, the rod is attached at the top vertex of the triangle, with top edges of both blocks in alignment.

- Draw and color the first eight terms of Sequence #2 by tracing the Cuisenaire Rods and Tangram pieces on plain paper.

- Compare your drawings with those of another group. If they differ, discuss and resolve your differences.

Thinking and Sharing

Have students post their drawings and compare and discuss any differences. Invite volunteers to explain how they went about building and extending the two sequences.

Use prompts like these to promote class discussion:

- How did you use the four given terms of Sequence #1 to help you figure out how to build the next six terms?
- How would you describe the pattern in Sequence #1?
- What mathematical clues did you find useful in building Sequence #2?
- What was easy to understand in the rules? What was hard? Explain.
- Did you and your partner interpret Sequence #2 in a different way than the group you compared with did? If so, what happened when you tried to resolve your differences?
- Which sequence (#1 or #2) was easier for you to work with? Explain.

On Their Own (Part 2)

What if... you and your partner were asked to design a sequence using both Cuisenaire Rods and Tangram pieces? What would your sequence be like?

- Work so that other groups cannot see the sequence you are designing.
- Using both Cuisenaire Rods and Tangram pieces, design and build the first six terms of an original sequence. The patterns in your sequence can be based on color, shape, position, number of pieces, or a combination of these and other characteristics.

- Draw the six terms of your sequence on plain paper. This will be the answer key.

- On another sheet of paper, write a set of rules that can be used to generate the terms of your sequence. Be as mathematically concise as possible.

- Exchange your rules with those of another group. Try to build the first six terms of the sequence described by the other group's rules.

- When both groups have finished, check the sequences against the answer keys. Discuss any differences until agreement is reached.

Thinking and Sharing

Ask students to explain how they went about designing their sequences and what happened when they tried writing rules describing them. Invite volunteers to display the drawings of their sequences and discuss any aspects of their design that they had difficulty describing.

Use prompts like these to promote class discussion:

- What was hard about designing a sequence? What was easy?

- What was hard about writing the set of rules? What was easy?

- How did you go about building the other group's sequence from their list of rules?

- Did you and your partner build a sequence that was different from the one intended by the other group? If so, why did this happen? Were both sequences valid interpretations of the rules? Explain.

- Looking back, what changes would you make in the way you described your sequence? Why would you make these changes?

For Their Portfolio

Think about the kinds of rules you wrote for your sequence. Then give an example of a rule that, although it is mathematically accurate, could be interpreted to generate two (or more) different sequences. Discuss the possible interpretations of your rule and ways to make it more precise.

Teacher Talk

Where's the Mathematics?

In this lesson, students consider sequence patterns that do not necessarily rely on numerical relationships. These kinds of patterns, which often appear as motifs on fabrics, carpets, and wallpapers, for example, illustrate sequences where numbers do not play a role in determining how the design is generated.

The terms generated by the sequences in the first activity are shown below.

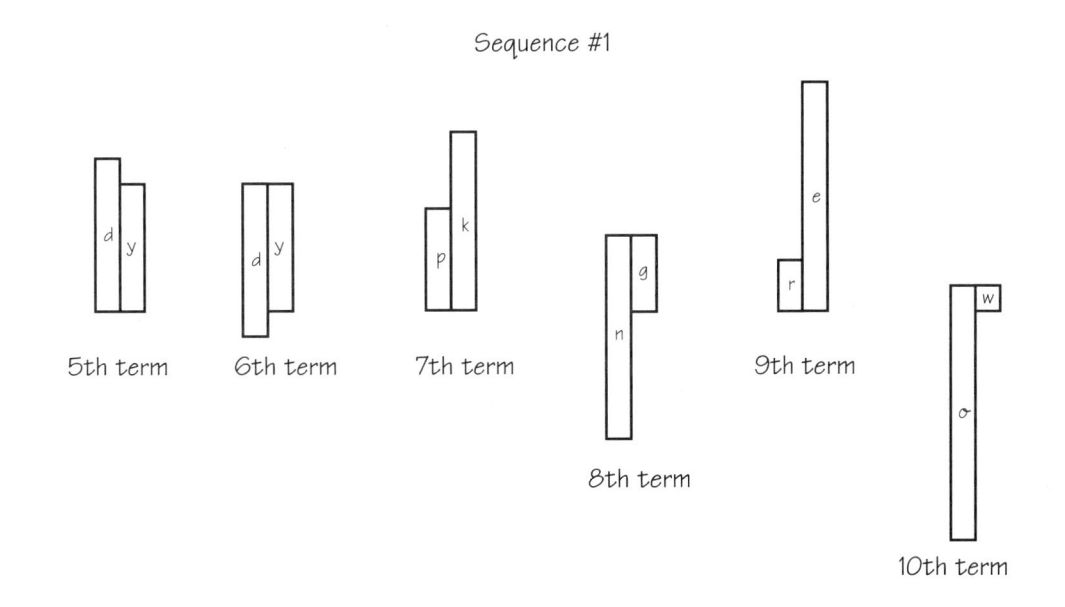

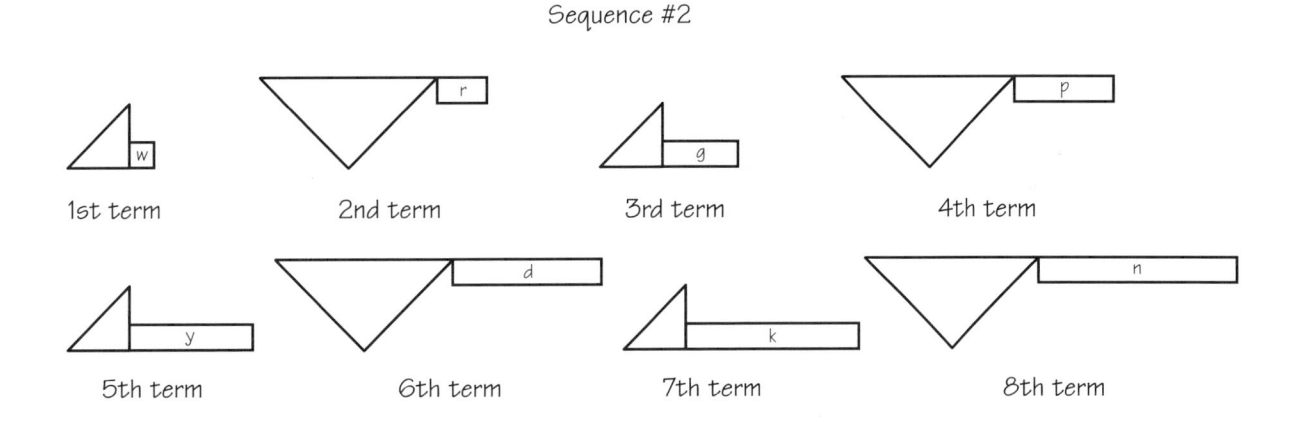

Sequence #1

5th term   6th term   7th term   8th term   9th term   10th term

Sequence #2

1st term   2nd term   3rd term   4th term

5th term   6th term   7th term   8th term

As students examine, create, and extend visual design sequences, they investigate patterns based on physical properties such as color, shape, arrangement, and position. The sequences they build may also involve patterns in the number of pieces used. Even students who are successful at designing sequences may encounter difficulty in explaining the rules, relationships, or connections among the terms. If this is the case, students may benefit by concentrating on just two or three consecutive terms, looking at their similarities and differences. This strategy can prevent students from becoming overwhelmed and can help them focus on the rule(s) that can be used to generate the terms of the sequence.

Searching for correct vocabulary to explain the way in which a sequence is generated helps students develop writing and communication skills in mathematics. As they write descriptions of what they have designed, students may combine visual and numerical observations in their rules. They may also recognize the need for more precise language.

©Cuisenaire Company of America, Inc.   VISUAL EFFECTS ♦ Patterns/Functions ♦ Grades 7-8

Some students may use more sophisticated mathematical terminology to explain the rules for their sequences. When working with Cuisenaire Rods and Tangram pieces, students may find that terms such as leg, hypotenuse, rotation, adjacent, edge, and vertex help to clarify their descriptions. They may also specify a relationship between the term number and the arrangement of pieces in the term.

Activities in which students create their own sequence designs and use them to challenge others underscore the importance of clarity in defining problems and in communication. Nowhere is this more evident than when two different valid interpretations of a set of rules leads to the generation of different sequences. Be sure to seize such an opportunity to highlight the importance of refining these skills should it occur in your classroom.

PASCAL PASTIMES

- Collecting and organizing data
- Pattern recognition
- Pascal's Triangle
- Experimental and theoretical probability

Getting Ready

What You'll Need

Geoboards, 1 per student

Rubber bands

Small sticky dots

Geodot paper, page 117

Color Tiles, 6 per pair

Activity master, page 114

Overview

Students search for all possible paths that can be made from a corner peg on a Geoboard to each of the other pegs. They then perform a probability experiment using Color Tiles and investigate how the outcome is related to the results of the first activity. In this activity, students have the opportunity to:

- analyze data to find patterns and make predictions
- relate visual patterns to numerical patterns
- learn about Pascal's Triangle and some of its applications
- compare the actual results of a probability experiment with the expected results

Other *Super Source* activities that explore these and related concepts are:

Visual Effects, page 78

Beehive Buzz, page 88

Carol's Kite Kits, page 93

The Activity

On Their Own (Part 1)

Eugene loves to play pinball and believes there is a way to figure out how many paths the silver ball could follow as it wends its way down through the bumpers. Can you help Eugene explore a method for finding the number of different paths the ball could take?

- Working with a partner, use the Geoboard to represent the pinball machine, its pegs to represent the bumpers, and rubber bands to represent the path of the ball. Position the Geoboard with a corner at the top and use lettered sticky dots to identify bumpers.

- Find all the paths the pinball could take as it travels from bumper A to bumper B, from A to C, from A to D, and so on, according to the following rule: As the ball moves from one peg (or bumper) to the next, it must travel downward to one of the two pegs closest to it in the next row.

©Cuisenaire Company of America, Inc.

- Look for all possible paths to each point. Keep track of the paths you find. (Geodot paper may be helpful.) Look for patterns that may help you predict the number of paths to particular pegs.

- On a clean sheet of geodot paper, record the *number* of paths you find from A to each bumper.

- Use your patterns to predict the number of different paths from A to Y without actually making them on your Geoboard. Be ready to explain your reasoning.

Thinking and Sharing

Draw (or project) a large Geoboard on the chalkboard, oriented and labeled as in *On Their Own*. Ask a volunteer to tell the number of different paths from A to B. Write this number over peg B. Then ask another volunteer to tell the number of paths from A to C, recording this number over peg C. Repeat this process for each peg. If students have different results, have them work together to reach agreement.

Use prompts like these to promote class discussion:

- How did you begin your search for various paths?
- How did you keep track of the paths of the pinball?
- How were you sure you did not miss or repeat a path?
- Did you notice any patterns while you were working? If so, describe them.
- Were you able to predict how many paths there might be to any particular peg(s)? If so, how?
- For which pegs was the number of paths easiest to predict? For which pegs was it more difficult?
- Looking at all of the data, what other patterns can you find?
- What predictions can you make about how the patterns might continue?

Explain to students that the data they collected produce part of a pattern known as *Pascal's Triangle* (see *Where's the Mathematics?*), a pattern that has many interesting mathematical applications.

On Their Own (Part 2)

What if... *Eugene is also curious about a game where pennies are tossed a certain number of times and the numbers of heads and tails thrown determines the winner? He might wonder if the outcomes follow a predictable pattern. What can you help Eugene discover about this game?*

- Working with your partner, place sticky dots on both sides of each of 4 Color Tiles. Let the tiles represent pennies. On each "penny," mark an "H" (heads) on one side and a "T" (tails) on the other.

- Have one partner act as recorder while the other partner tosses the tiles.

- Toss all 4 marked Color Tiles at the same time onto a flat surface. Count the number of "heads" and "tails" and record the outcome with a tally mark in a chart like the one shown.

Heads:	4	3	2	1	0
Tails:	0	1	2	3	4
Tally:					
Totals:					

- Toss the tiles 15 more times, recording each outcome in the chart.
- Calculate the totals and look for patterns.
- Conduct similar experiments, tossing 5 marked Color Tiles 32 times and then 6 marked Color Tiles 64 times. Record the outcomes in charts like the one above.
- Discuss your results with your partner. What patterns do you see? What predictions can you make? What relationships do you see between the outcomes of this problem and the pinball path problem?

Thinking and Sharing

Have students share their results and observations. Create a class chart for each part of the experiment, listing the findings of each group.

Use prompts like these to promote class discussion:

- What did you notice about the outcomes of your experiments?
- Were you surprised by your results? Explain.
- What do you notice about the data collected by the different groups?
- What would you expect to happen if you repeated the experiments?
- What patterns do you see in the data?
- How does the data collected relate to Pascal's Triangle?

If the distribution of outcomes recorded by each pair of students does not approximate the expected values predicted by Pascal's Triangle, you may want to combine the class data in each chart and have students compare the ratios of the totals to the ratios of the numbers in the applicable rows of Pascal's Triangle (see *Where's the Mathematics?*).

For Their Portfolio

Investigate other applications of Pascal's Triangle and write a few paragraphs describing one application that you find particularly interesting.

Teacher Talk

Where's the Mathematics?

There are 250 different paths from peg A to the other 24 pegs on the Geoboard. Thus, as students work through the activity, they may find it challenging to keep track of the numerous paths traveled by the pinball. Some students may draw each of the paths they find on geodot paper while others may try using different-colored rubber bands and several Geoboards. Some groups may develop a systematic approach to searching for all possible paths to a specific peg, perhaps assigning different tasks to various group members.

Some students may search randomly and compare their results to those found by other group members, making sure duplications have been eliminated. Students may choose to first investigate all paths in which the first move in each path is downward to the left and then to work on those in which the first move is downward to the right. Other students may design some type of recording system based on the sequencing of letter arrangements. Here is one possible way of recording paths from peg A to each of pegs K, L, and M.

A → K	A → L	A → M
ABDGK	ABDGL	ABDHM
	ABDHL	ABEHM
1 PATH	ABEHL	ABEIM
	ACEHL	ACEHM
		ACEIM
	4 PATHS	ACFIM
		6 PATHS

If students are familiar with tree diagrams, they may choose to use them to record paths. An example of a tree diagram for paths from peg A to peg M is shown below:

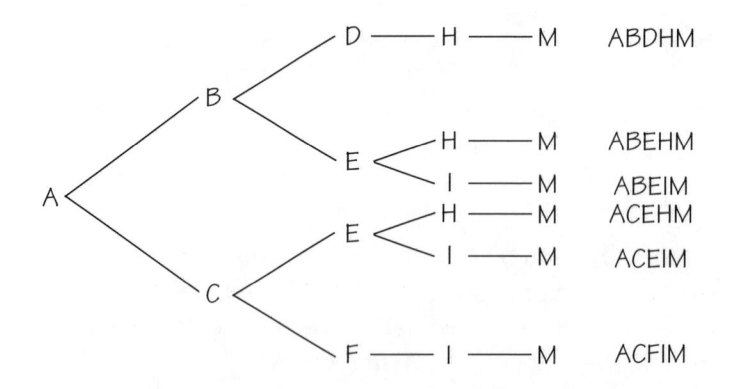

```
              D ── H ── M     ABDHM
         B <
              E < H ── M      ABEHM
A <               I ── M      ABEIM
              E < H ── M      ACEHM
         C <      I ── M      ACEIM
              F ── I ── M     ACFIM
```

As students continue to search for paths, they may look for patterns in the data they have collected. They may also find that the patterns they identify may help them in confirming the data they have already gathered. The number of paths from peg A to each of the other bumper pegs is shown at right.

```
           •A
        •1    •1
     •1    •2    •1
  •1    •3    •3    •1
•1   •4    •6    •4   •1
   •5   •10   •10   •5
      •15   •20   15•
         •35   35•
            •70
```

Students should discover a number of different patterns in the data. They may notice that only one path is possible to each of the pegs along the upper left and right edges of the Geoboard. They may also notice that the data form a symmetrical pattern, or *palindromic* pattern, meaning that they read the same from left to right as they do from right to left. Some students may realize that the number of paths to a particular peg is the sum of the numbers of paths to the pegs immediately above it to the left and to the right. For example, if there are 6 paths from peg A to peg M and 4 paths from peg A to peg N, there will be 6 + 4, or 10, paths from peg A to peg R. Students can use these patterns to predict the number of paths to any peg on the Geoboard.

Looking at the completed work, students will see part of a pattern known as *Pascal's Triangle*, the first seven rows of which are shown below.

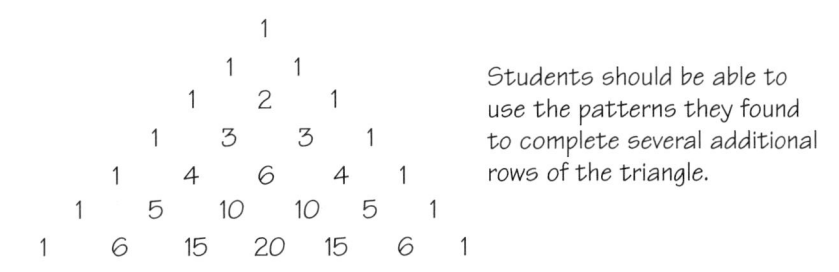

```
            1
          1   1
        1   2   1
      1   3   3   1
    1   4   6   4   1
  1   5  10  10   5   1
1   6  15  20  15   6   1
```

Students should be able to use the patterns they found to complete several additional rows of the triangle.

As mentioned, there are many applications of Pascal's Triangle. The second activity illustrates one such application. The probability experiments illustrate how Pascal's Triangle can be used to predict the outcomes of certain types of events. Since there are only two possible ways for each tile to land ("heads" or "tails"), and the tiles are fair, the chance is the same for either one of the two outcomes to occur. Below are charts for each of the three experiments, with data illustrating the hypothetical results (the number of times each outcome can be *expected* to occur) based on rows from Pascal's Triangle.

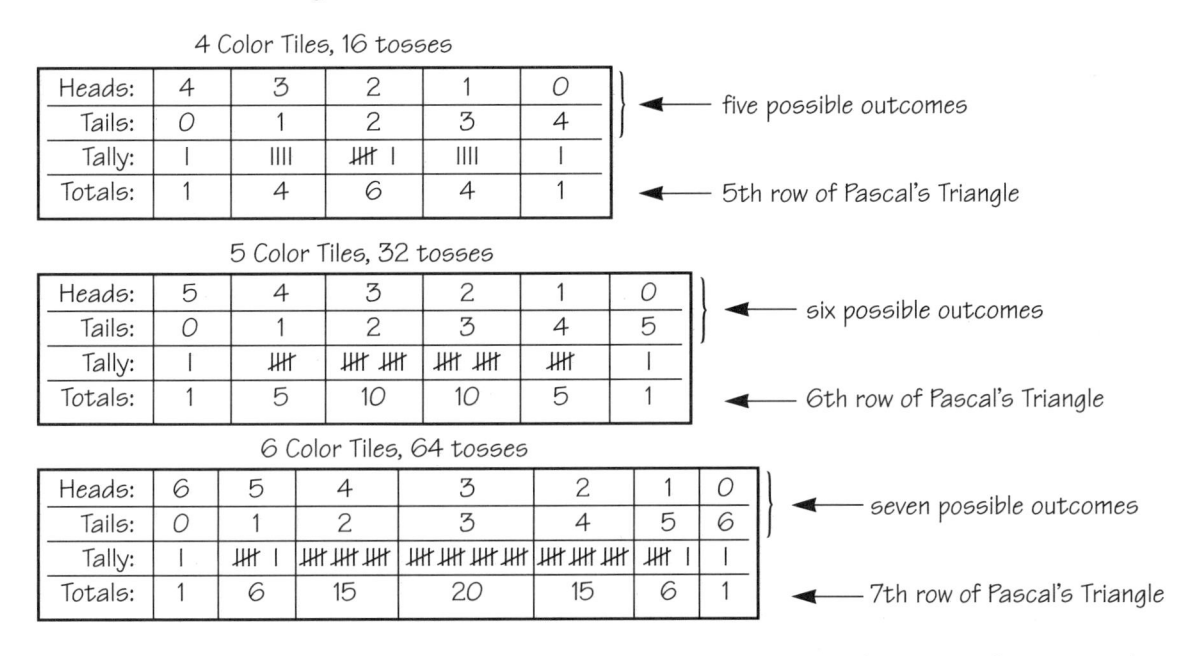

4 Color Tiles, 16 tosses

Heads:	4	3	2	1	0
Tails:	0	1	2	3	4
Tally:	I	IIII	HHT I	IIII	I
Totals:	1	4	6	4	1

← five possible outcomes

← 5th row of Pascal's Triangle

5 Color Tiles, 32 tosses

Heads:	5	4	3	2	1	0
Tails:	0	1	2	3	4	5
Tally:	I	HHT	HHT HHT	HHT HHT	HHT	I
Totals:	1	5	10	10	5	1

← six possible outcomes

← 6th row of Pascal's Triangle

6 Color Tiles, 64 tosses

Heads:	6	5	4	3	2	1	0
Tails:	0	1	2	3	4	5	6
Tally:	I	HHT I	HHT HHT HHT	HHT HHT HHT HHT	HHT HHT HHT	HHT I	I
Totals:	1	6	15	20	15	6	1

← seven possible outcomes

← 7th row of Pascal's Triangle

From their data, students can generalize that when n pennies are tossed 2^n times, the expected outcomes can be predicted by the numbers in the $(n + 1)$th row of Pascal's Triangle. Students may enjoy comparing their results to those predicted by Pascal's Triangle. Their results may be different from those expected because of the limited number of tosses or samples. However, if groups combine their data, they are apt to find that the ratios of their outcomes to those predicted are more in agreement.

BEEHIVE BUZZ

- Equivalence
- Pattern recognition
- Organizing data

Getting Ready

What You'll Need

Pattern Blocks, at least 14 hexagons, 14 trapezoids, 21 blue rhombuses, and 84 triangles per pair

Activity master, page 115

Overview

Students search to find combinations of Pattern Blocks that can be used to cover a honeycomb design made from 7 hexagons. In this activity, students have the opportunity to:

- work with geometric equivalence
- collect and organize data
- use patterns to make predictions
- investigate sequences generated by geometric relationships

Other *Super Source* activities that explore these and related concepts are:

Visual Effects, page 78

Pascal Pastimes, page 83

Carol's Kite Kits, page 93

The Activity

On Their Own (Part 1)

A beehive is abuzz with activity as its bee colony, consisting of queen bees, drones, and workers, moves in to make honey in a 7-cell honeycomb. If different-shaped Pattern Blocks are used to represent the honeycomb and the three types of bees, how many different combinations of bees can completely occupy the honeycomb?

- Work with a partner. One of you should work on Honeycomb A while the other works on Honeycomb B.

- Each of you should begin by placing 7 yellow hexagon blocks together to make the honeycomb shown at the right.

Honeycomb A
- Your honeycomb will be occupied by queen bees (red trapezoids) and workers (green triangles).

Honeycomb B
- Your honeycomb will be occupied by drones (blue rhombuses) and workers (green triangles).

- Use only red and green Pattern Blocks to cover your hexagon honeycomb.

- Record the number of red queens, the number of green workers, and the total number of bees in your honeycomb.

- Now use different combinations of your Pattern Block bees to cover your hexagon honeycomb. See how many different combinations you can find. Organize your findings in a table.

- Use only blue and green Pattern Blocks to cover your hexagon honeycomb.

- Record the number of blue drones, the number of green workers, and the total number of bees in your honeycomb.

- Now use different combinations of your Pattern Block bees to cover your hexagon honeycomb. See how many different combinations you can find. Organize your findings in a table.

- Share and compare your findings with those of your partner. Look for relationships among the numbers and types of bees occupying the honeycombs.

Thinking and Sharing

Invite several students who worked on Honeycomb A to make a chart displaying their results on the board. Do the same for Honeycomb B. If students who worked on the same honeycomb have data that differ, have them work together to come to agreement.

Use prompts like these to promote class discussion:

- How many different combinations of red and green blocks did you find?
- How many different combinations of blue and green blocks did you find?
- What strategies did you use for solving the problem?
- Did you change strategies as you worked? If so, how did you change, and why?
- Did you notice any patterns that helped you find solutions? If so, describe them.
- What relationships did you find among the blocks that you used and the combinations you found?
- Did you notice any similarities and/or differences between your results and those of your partner? If so, describe them.

On Their Own (Part 2)

What if... the 7-cell honeycomb is occupied by queens, drones, and workers? What combinations of bees might be found in the honeycomb?

- Working with your partner, build one 7-cell honeycomb using yellow hexagons.

- Using red trapezoids (queen bees), blue rhombuses (drones), and green triangles (workers), find a way to cover your hexagon honeycomb.

- Record the number of red queens, blue drones, and green workers, and the total number of bees in your honeycomb.

- Now use different combinations of Pattern Block bees to cover your hexagon honeycomb. See how many different combinations you can find. Organize your findings in a table.

- Look for relationships among the numbers and types of bees occupying the honeycomb.

- Predict the maximum number of queens, drones, and workers that could occupy a honeycomb with n cells. Be ready to justify your conclusions.

Thinking and Sharing

Ask students to help you create a class chart displaying the data. Have them help you decide the best way to organize and arrange the data in the table so that the patterns are easy to see.

Use prompts like these to promote class discussion:

- Was this activity harder or easier than the first activity? Explain why.
- How many different combinations of red, blue, and green blocks did you find?
- What was the largest number of blocks you used to fill your honeycomb? What was the smallest? Do you think there are ways that they can be filled with larger or smaller numbers of blocks? Explain.
- What strategies did you use for finding combinations?
- Did you change strategies as you worked? If so, how did you change, and why?
- Did you notice any patterns that helped you find solutions? If so, describe them.
- What predictions did you make about the maximum number of queens, drones, and workers that could occupy a honeycomb with n cells? Explain your thinking.

For Their Portfolio

Write an explanation of the strategies you used to organize your data as you searched for solutions to the honeycomb problems.

Teacher Talk

Where's the Mathematics?

Students may investigate the problems in this activity using a variety of methods. Some may use trial and error. Others may begin by trial and error, and then develop a more systematic approach. As they start to compile data, students will find that many different combinations of Pattern Blocks can be used to fill the 7-cell honeycomb, and that they must either do a lot of counting, or search for patterns that can be used to predict solutions.

Some students may begin their search by covering the 7 hexagons with the largest Pattern Blocks they are working with. In Honeycomb A, 14 red trapezoids are required to cover the honeycomb, while in Honeycomb B, 21 blue rhombuses are required. Other students may begin with the smallest Pattern Blocks, determining that in both Honeycomb A and Honeycomb B, 42 green triangles are needed to cover the 7 hexagons.

Students should be aware of the equivalence relations among blocks as they look for additional solutions: One yellow hexagon is equivalent to 2 red trapezoids, or 3 blue rhombuses, or 6 green triangles. One red trapezoid is equivalent to 3 green triangles. One blue rhombus is equivalent to 2 green triangles. Substitutions based on these equivalences will make it easier for students to generate new solutions from existing ones. After using this substitution strategy several times, students may be able to identify patterns that may help them organize their data and determine additional solutions without actually working with the blocks. Students should be encouraged to verify some of these predictions for accuracy.

The charts below show solutions for each of the honeycombs.

Honeycomb A

number of red queens	number of green workers	total number of bees
14	0	14
13	3	16
12	6	18
11	9	20
10	12	22
9	15	24
8	18	26
7	21	28
6	24	30
5	27	32
4	30	34
3	33	36
2	36	38
1	39	40
0	42	42

Honeycomb B

number of blue drones	number of green workers	total number of bees
21	0	21
20	2	22
19	4	23
18	6	24
17	8	25
16	10	26
15	12	27
14	14	28
13	16	29
12	18	30
11	20	31
10	22	32
9	24	33
8	26	34
7	28	35
6	30	36
5	32	37
4	34	38
3	36	39
2	38	40
1	40	41
0	42	42

Some students may point out that the honeycombs were to be occupied by both queens and workers, or by both drones and workers, and may argue that therefore the first and last "solutions" listed for each honeycomb are not valid.

When students describe the patterns they observed, they are apt to point out that as the number of red blocks in Honeycomb A decreases by 1, the number of green blocks increases by 3. They may also notice that the numbers in the total column increase by 2 and are always even integers. Students may be able to explain that this happens because every time 1 red trapezoid is exchanged for 3 green triangles, there are two more pieces on the 7-cell honeycomb. For Honeycomb B, as the number of blue blocks decreases by 1, the number of green blocks increases by 2. They may also notice that the numbers in the total column increase by 1. Using reasoning similar to that above, students may explain that this happens because every time 1 blue rhombus is exchanged for 2 green triangles, there is one more piece on the 7-cell honeycomb.

The second activity presents a challenge in that students must organize a considerable quantity of data. If they have found a good way to organize their data in the first part of the activity, they may have a head start on approaching the problem involving three different-sized blocks.

Students may choose to start with the maximum number of red trapezoids needed to cover the 7 hexagons (14) and then decrease this number one trapezoid at a time. Each trapezoid removed could be replaced with either 1 blue rhombus and 1 green triangle, or 3 green triangles to generate new combinations. Using these substitution strategies, students can systematically look for combinations and organize the data. A portion of the data, organized as described, is shown below.

The patterns students find and point out will vary depending on how they organized their data. In the chart shown here, students may notice clusters of decreasing consecutive integers in the second column, clusters of increasing odds and increasing evens in the third column, and clusters of increasing integers in the fourth column. They may also point out that the number of terms in each cluster increases as the number of red queens decreases.

As students work back and forth between the blocks and the data, they may discover a way to use the emerging patterns to generate data for new combinations without actually assembling them. They may also recognize how the numbers of "bee blocks" are related to the number of cells in a hexagon honeycomb.

number of red queens	number of blue drones	number of green workers	total number of bees
14	0	0	14
13	1	1	15
13	0	3	16
12	3	0	15
12	2	2	16
12	1	4	17
12	0	6	18
11	4	1	16
11	3	3	17
11	2	5	18
11	1	7	19
11	0	9	20
10	6	0	16
10	5	2	17
10	4	4	18
10	3	6	19
⋮	⋮	⋮	⋮

Students should conclude that if a hexagon honeycomb structure contains n cells, the maximum number of red trapezoids needed to cover the surface is $2n$, the maximum number of blue rhombuses needed is $3n$, and the maximum number of green triangles is $6n$. They should also come to realize that these generalizations would enable them to find combinations of bees for any size honeycomb without having to work with the blocks.

CAROL'S KITE KITS

- Solid geometry
- Using patterns
- Area
- Writing algebraic expressions

Getting Ready

What You'll Need

Snap Cubes, about 130 per pair

Isometric dot paper, page 123

1-centimeter grid paper, page 121

Activity master, page 116

Overview

Students use Snap Cubes to build increasingly larger models of kite frames. They gather data about the amount of materials needed to build each kite and look for underlying number patterns. In this activity, students have the opportunity to:

- build and draw 3-dimensional models
- collect, organize, and analyze data
- use patterns to make predictions
- review formulas for finding the areas of rectangles and triangles
- use algebraic expressions to generalize patterns

Other *Super Source* activities that explore these and related concepts are:

Visual Effects, page 78

Pascal Pastimes, page 83

Beehive Buzz, page 88

The Activity

On Their Own (Part 1)

Carol designs kite kits of various sizes and styles. One of the most popular styles is the box kite. The frame is constructed of lightweight wooden strips with brightly colored bands of paper attached to the top and bottom of the frame. Can you help Carol determine the amounts of wood and paper needed for different-sized box kite kits?

- Working with a partner, use 2 Snap Cubes to build a model of kite #1 with dimensions 1 x 1 x 2. Let the edges of the structure represent the frame of the kite, and draw a model of this kite on isometric dot paper.
- Determine and record the total length of wooden strips needed to build the frame of kite #1. Let the edge of a cube represent 1 unit.

frame

(¼ height)

(½ height)

(¼ height)

©Cuisenaire Company of America, Inc.   CAROL'S KITE KITS ◆ Patterns/Functions ◆ Grades 7-8   **93**

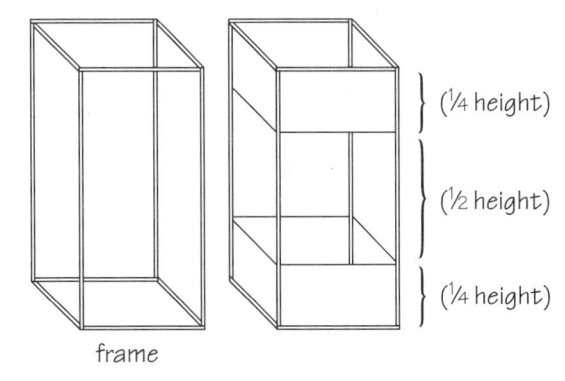

- Two bands of paper are needed to finish the kite's construction. Each band is one fourth the height of the frame. Determine and record the total amount of paper (in square units) needed for kite #1.
- Build kite models with dimensions 2 x 2 x 4, 3 x 3 x 6, and so on, where the height is twice the side length of the square base. Draw each model on dot paper and record the amounts of wood and paper needed to build it.
- Organize your data in a chart. Look for patterns to help you complete the data for the first six kite kits. Then predict the amounts of materials needed to build a kite measuring 10 x 10 x 20.
- Write algebraic expressions that generalize the patterns you find. Be ready to explain your findings.

Thinking and Sharing

Create a class chart with column headings: *kite #, dimensions of base, height, total length of wood needed,* and *amount of paper needed*. Have students help you complete the chart with their data for the first six kites and their predictions for the tenth kite.

Use prompts like these to promote class discussion:

- What did you notice as you built successively larger kite models?
- How did you go about drawing the models on isometric dot paper?
- What patterns and/or sequences did you discover in the data?
- How did you calculate the amounts of wood and paper needed for each of the kites?
- What strategies did you use to predict the materials needed for a 10 x 10 x 20 kite?
- What algebraic expressions did you write to represent the patterns and/or relationships you found.

On Their Own (Part 2)

What if... Carol decides to make kits for diamond-shaped kites? If the frame of each kite is made from two wooden strips and is covered with a thin layer of paper, how much wood and paper would be needed for each kit?

- Working with a partner, use Snap Cubes to build a model of the frame for the first kite as follows:
  - Start with one cube representing the portion of the frame where the wooden strips overlap.

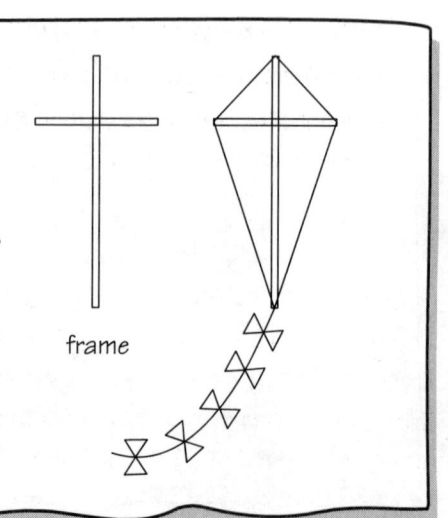

frame

- ◆ Attach one cube to the top face of the cube, one to its left face, one to its right face, and two to its bottom face.

- Draw your model of this kite frame on grid paper and label it kite #1. Determine and record the lengths of each of the two pieces of wood needed to build the frame, as well as the total amount of wood needed. Let the edge of a cube represent 1 unit.

- Determine and record the total amount of paper (in square units) needed to cover the frame of kite #1.

- Build models of other diamond-shaped kite frames where the number of cubes attached to the bottom face of the overlap cube is twice the number of cubes attached to the top, left, and right faces. Draw and number each model on grid paper and record the amounts of wood and paper needed to build each kite.

- Organize your data in a chart. Look for patterns.

- Write algebraic expressions that generalize the patterns you find. Be ready to explain your findings.

Thinking and Sharing

Create a class chart with column headings: *kite #, length of horizontal strip, length of vertical strip, total length of wood needed,* and *amount of paper needed.* Ask students to help you complete the chart with their data. Have them work together to resolve any differing measurements.

Use prompts like these to promote class discussion:

- What did you notice as you built successively larger kite models?
- How did you find the total amount of wood needed for each kite?
- How did you find the amount of paper needed for each kite?
- What patterns and/or sequences did you discover in the data?
- What algebraic expressions did you write to represent the patterns and/or relationships you found?

For Their Portfolio

Design your own kite. Draw a picture of it on isometric dot paper, grid paper, or plain paper. Then write a letter to Carol, asking her to consider making and selling kits for your kite. Describe how your kite is constructed, and explain ways to determine the amounts of materials needed for kits of different sizes.

Teacher Talk

Where's the Mathematics?

As students build models of increasingly larger box kite frames and determine the amounts of materials needed to build the kites, they can begin to see relationships among the "kite number," its dimensions, and the amounts of materials needed for the kit. The chart helps to bring out the emerging patterns.

Box Kite Kits

kite #	base	height	total length of wood needed	amount of paper needed
1	1 x 1	2	16	4
2	2 x 2	4	32	16
3	3 x 3	6	48	36
4	4 x 4	8	64	64
5	5 x 5	10	80	100
6	6 x 6	12	96	144
⋮	⋮	⋮	⋮	⋮
10	10 x 10	20	160	400
⋮	⋮	⋮	⋮	⋮
n	n x n	$2n$	$16n$	$4n^2$

Students should note that the bases of successive box kites are squares whose side lengths are increasing consecutive integers. The side length is also the same as the kite number. They can therefore predict that the base of the 10th kite will be a 10 x 10 square, and the base of the nth kite will be an n x n square. The entries in the *height* column are each twice the kite number (or the length of a side of the base) and, therefore, can be represented in general by $2n$.

In studying the data for the total length of wood needed for the box kite's frame, students may observe that the amounts differ by 16, forming an arithmetic sequence. It is also true that the total length of wood needed for the frame is 16 times the kite number (or the length of a side of the base). To predict the total length of wood needed for the 10 x 10 x 20 box kite frame, students can either continue the pattern to find the tenth entry in the column, or multiply the side length, 10, by 16 to get 160. Similarly, the total length of wood needed for the nth kite is $16n$ (8 strips of length n, plus 4 strips of length $2n$).

Students may consider each of the paper bands to be a rectangle whose length is the perimeter of the square base. The height of each rectangle is one-fourth the height of the frame, as specified. Using the formula for area of a rectangle, students can find that the amount of paper for one band can be found using the formula: ¼ x height x perimeter of base. Thus, for the 10th kite, the area of each band is ¼ x 20 x 40, or 200 square units, and the total amount of paper needed would be 400 square units. In general, the total amount of paper needed for the nth kite is 2(¼ x 2n x 4n), or $4n^2$, square units.

Some students may point out that the entries in the last column are perfect square numbers. They may be intrigued to discover that the terms generated by the expression $4n^2$ are the squares of consecutive even integers.

The chart on the next page shows the data for for the diamond-shaped kite.

Diamond-shaped Kite Kits

kite #	length of horizontal strip	length of vertical strip	total length of wood needed	amount of paper needed
1	3	4	7	6
2	5	7	12	17.5
3	7	10	17	35
4	9	13	22	58.5
5	11	16	27	88
6	13	19	32	123.5
⋮	⋮	⋮	⋮	⋮
n	$2n + 1$	$3n + 1$	$5n + 2$	$\frac{1}{2}(6n^2 + 5n + 1)$

Students may notice that the entries in the *length of horizontal strip* column are consecutive odd integers, each 1 more than twice the kite number. The entries in the *length of vertical strip* column also form an arithmetic sequence and are each 1 more than 3 times the kite number. In general, if n represents the kite number, then the length of the horizontal strip is $2n + 1$, the length of the vertical strip is $3n + 1$, and the total length of the wood needed is $(2n + 1) + (3n + 1)$, or $5n + 2$, an expression which also generates an arithmetic sequence as evidenced by the entries in the fourth column.

Students may have different ways of figuring out the amount of paper needed to cover the diamond-shaped kite. Some may use the wood frame to visually separate the paper into 2 large triangles or 4 smaller right triangles. In either case, the lengths of the parts of the frame can then be used to calculate the area of each triangle using the area formula: Area = $\frac{1}{2}$ × base × height. For example, if students use the vertical strip to separate the paper into two congruent triangles with equal areas, they need only find the area of one of the two triangles, and then double that area to find the total surface area of the kite. Using the length of the vertical wood strip for the base and one half the length of the horizontal strip for the height, students can find the area of the triangle. For example, in the 4th kite, where the length of the vertical strip is 13 units and the length of the horizontal strip is 9 units, the area of one triangle is $\frac{1}{2}$ × 13 × 4.5, or 29.25 square units. Thus, the total area of the kite surface is 2 × 29.25, or 58.5 square units.

For the general case of the nth diamond-shaped kite, students can use the algebraic expressions they write for the lengths of the horizontal and vertical strips to determine information about the base and height of the triangular sections. If they think of the paper as two triangles separated by the vertical strip, the amount of paper needed can be expressed as $2[\frac{1}{2} \times (3n + 1) \times \frac{1}{2}(2n + 1)]$, or, when simplified, $\frac{1}{2}(6n^2 + 5n + 1)$ square units. Some students may instead reflect and rotate the two left-hand triangles onto the two right-hand triangles, creating a single rectangle with length equal to the height of the kite and width equal to half the width of the kite. Then, the amount of paper needed could be expressed as a product of the length and the width: $(3n + 1) \times \frac{1}{2}(2n + 1)$, or $\frac{1}{2}(6n^2 + 5n + 1)$ square units. Some students may have difficulty simplifying their algebraic expressions. You may want to allow them to leave the expressions in factored form.

As an intriguing footnote, you may want to ask students to explore the ratios of horizontal strip length to vertical strip length. In kite #1, that ratio is $\frac{3}{4}$, and as the kite number increases, the ratio steadily decreases, approaching a limit of $\frac{2}{3}$. Students can readily grasp this idea if they look at ratios like $\frac{21}{31}$ (at $n = 10$) compared to $\frac{201}{301}$ (at $n = 100$) or $\frac{2001}{3001}$ (at $n = 1000$).

Blackline Masters

Backyard Improvements

Part 1

Alan is planning to build a square patio in his backyard using square pieces of slate that measure 1 foot by 1 foot. He needs to decide what dimensions to make the patio so that he can determine how many pieces of slate to order. What information can you gather that might help Alan with his project?

- Use the orange Pattern Blocks to represent the square pieces of slate. Working with your partner, build models of increasingly larger square patios, beginning with the smallest possible patio, the one made from one square.
- Each time you build a new patio model, record the number of blocks you added to the previous model to build the next bigger square, the perimeter of the patio, and the total number of blocks in the new patio. Organize your data in a table.
- Record the data for the first ten squares, but build squares only until you discover a pattern that will produce all the numbers needed for your table.
- Look for relationships between the dimensions of the square patio, its perimeter, and the total number of pieces of slate needed. Think about how you might generalize your findings. For example, think about what the data would be for a patio with dimensions n feet by n feet.
- Be ready to discuss and justify your conclusions.

Part 2

What if... Alan also decides to build a set of stairs from one level of his backyard to another? He wants to use railroad ties for the steps, and must determine how many he needs. What information can you gather that might help Alan with this project?

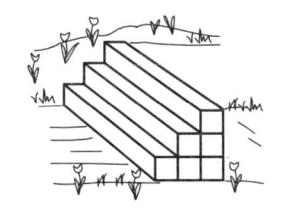

- Using the orange squares to represent the side view of the railroad ties, work with your partner to build models of increasingly larger staircases.
- Each time you build a new staircase, record the number of blocks you added to build the next taller step, the perimeter of the new side view, and the total number of ties needed to build the staircase. Organize your data in a table.
- Record the data for the first ten staircases, but build sets of stairs only until you discover a pattern that will produce the numbers needed for your table.
- Look for relationships between the number of steps in the staircase, the perimeter of the side view, and the total number of railroad ties needed. Think about how you might generalize your findings to a staircase with n steps.
- Compare the sequences you found in the first activity with the sequences you found in this activity. Discuss your observations with your partner.

For Your Portfolio

Describe how the various sequences you found in these activities were alike and how they were different. Make up some sequences of your own that are similar in some way to the ones that were generated by the data.

Ripples

Part 1

Mariano notices that when a stone is thrown into a calm body of water, it produces a ripple effect of larger and larger circles whose centers are the stone's point of contact with the water's surface. Mariano wants to investigate a similar kind of ripple effect using Pattern Block shapes. What patterns can be discovered in a sequence of ripples?

- Work with a partner. Using the Pattern Block square or the blue or tan rhombus as the "stone," completely surround your stone with blocks of the same shape to form the first "ripple." Record your design. Color it using one color for the original stone and another color for the blocks in the first ripple.
- Surround your first ripple design with more blocks of the same shape to form the second ripple. Make sure that the entire perimeter of the ripple is surrounded with new blocks. Record your new design. Color the new ripple with a different color.
- Predict how many stones it will take to form the third ripple and check your prediction.
- As each new ripple is generated, record the number of blocks in the new ripple, the total number of blocks in the ripple design, and the perimeter of the ripple design.
- Continue predicting, surrounding your designs, and recording the data until you have built the sixth ripple.
- Look for patterns in your results. See if you can write algebraic expressions that generalize your findings.
- Repeat the activity using yellow Pattern Block hexagons.

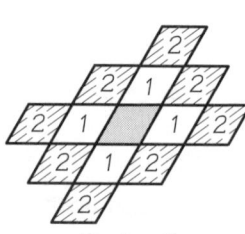

Design 1

Design 2

Part 2

What if... Mariano decides to investigate the ripple patterns for the triangle and trapezoid stones? Will he find that these shapes produce patterns similar to the ones he found for the other shapes?

- Using the procedure from the first part of the activity, gather data about ripples formed from using the Pattern Block triangle as the stone.
- Compare the patterns formed by the triangle data with those you found in Part 1.
- Predict what will happen if you use the Pattern Block trapezoid for the stone. Test your prediction by building the first three ripple designs. As you add trapezoids to the design, be sure to join short edge to short edge and long edge to long edge.
- Discuss whether or not the trapezoid designs follow growth patterns that are similar to those of the other shapes. Be ready to justify your findings.

For Your Portfolio

Write a letter to Mariano explaining how the shape of a stone affects the patterns formed by its ripples. Use examples and drawings to help illustrate your explanation.

Marquetry

Part 1

Marquetry is a technique in which pieces of different types of wood are inlaid into the surface of a piece of furniture to create a design. One popular design often used to decorate table tops is the Greek Cross design made from cubes of wood of varying colors. This type of design is not only interesting to look at, but is also full of hidden patterns. Can you find them?

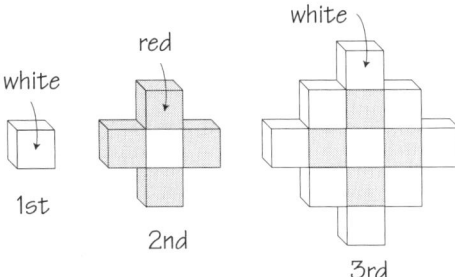

- Using Snap Cubes to represent the cubes of wood, work with a partner to build the Greek Cross designs shown. Use only two colors (white and red, for example) for your designs, alternating colors as you build each larger design.
- For each Greek Cross design, record the design number, the number of cubes added to the previous structure, the number of cubes of each color, and the total number of cubes in the structure. Organize your data in a chart.
- Predict the number of cubes of each color that would be needed to build the fourth Greek Cross structure and check your prediction by building it. Do the same for the fifth structure.
- Look for patterns in your data. Use the patterns to continue your chart through the 10th Greek Cross design. Then generalize your findings by writing algebraic expressions for the data for the *n*th Greek Cross design, where possible.

Part 2

What if... the corner areas of the Greek Cross designs are filled in with wooden cubes? How would this affect the numbers of the different-colored cubes needed for each design?

- Using your models from the first activity, fill in the open "corners" of each Greek Cross with Snap Cubes to create a square. In each case, use the color that would come next in the alternating sequence.
- For each marquetry design, determine the dimensions of the square, the number of cubes added at each corner, the total number (and color) of cubes added, and the total number of cubes in the square design. Add columns to the chart you made in the first activity to record your data.
- Look for patterns in your data. Use the patterns to continue your chart through the 10th design. Then generalize your findings by writing algebraic expressions for the data for the *n*th design.
- Compare the different sequences formed by the entries in you chart. Discuss them with your partner and be ready to talk about any relationships you find.

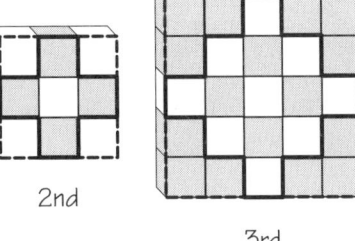

For Your Portfolio

Picture a 25-inch by 25-inch square marquetry design made from cubes of wood measuring 1 inch on each edge. If the design is to contain the largest Greek Cross that can be fitted in the square, what materials will be needed? Explain your reasoning and show your calculations.

©Cuisenaire Company of America, Inc.   Patterns/Functions ♦ Grades 7-8   **101**

The Pyramid Mystery

Part 1

The Egyptians were known for their magnificent granite and limestone pyramids and the remarkably mysterious methods used in building them. Can you build models of the pyramids and unlock the mystery surrounding the number of blocks needed for their construction?

- Using Snap Cubes to represent blocks of granite and limestone, work with a partner to build pyramids like those shown below. Use only two colors (yellow and brown, for example) for your pyramids. Alternate colors, so that each new outer layer of cubes is the opposite color of the layer underneath.

- For each structure, record its number in the sequence of pyramids, the number of cubes added to the previous structure, the number of cubes of each color, and the total number of cubes in the pyramid. Organize your data in a chart.

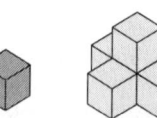

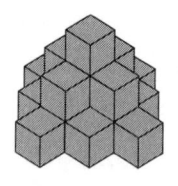

- Predict what cubes would be needed to build the fourth pyramid. Check your prediction by building the fourth pyramid.

- Look for patterns in your data. Use the patterns to continue your chart through the 10th pyramid.

- Compare the sequences formed by the entries in the different columns of your chart. Be ready to talk about your observations and any relationships you find.

Part 2

What if... the sequence of structures was generated by surrounding each previous structure on all sides with cubes of a contrasting color? How would this affect the numbers of different-colored cubes needed for each structure?

- Using Snap Cubes, work with a partner to build "octamids" like those shown below. Use the same color arrangements as you used in the first activity.

- For each structure, record its number in the sequence of octamids, the number of cubes in its outer shell, the number of cubes of each color, and the total number of cubes in the octamid. Organize your data in a chart.

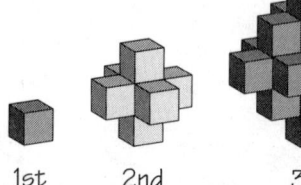

- Look for patterns in your data. Use the patterns to continue your chart through the 10th octamid.

- Using the charts you made in the two activities, look for patterns and relationships among the sequences formed by the data. Be ready to discuss your findings.

For Your Portfolio

Write a short paragraph or two describing the relationship between the Greek Cross numbers and the Pyramid numbers. Explain the relationship both geometrically and numerically, using diagrams where helpful to illustrate your explanations.

Bees in the Trees

Part 1

The family tree of a male bee is both unusual and interesting. The male bee is created through a process known as parthenogenesis, whereby he has a single parent, only a mother. The female bee, however, has both a mother and a father. What patterns can you find by tracing the ancestry of a male bee?

- Work with a partner. Construct a family tree of a male bee using black Snap Cubes to represent male bees, yellow Snap Cubes to represent female bees, and toothpicks to represent connections from generation to generation. Here's how:
    - Place a black Snap Cube on your work surface to represent the male bee. Place a toothpick "connecting" this cube to a yellow Snap Cube representing the male bee's mother.
    - From the mother bee, use 2 toothpicks — one leading to a black cube, the other to a yellow cube — to show the connection to her two parents.
    - Making sure to keep all ancestors of the same generation in the same horizontal row, continue the process of adding toothpicks and yellow and black cubes to trace back 8 complete generations in the male bee's family tree.
- As each generation is added, record the generation number, the number of male bees, the number of female bees, the number of bees in that generation, and the total number of ancestors. Organize your data in a chart.
- Look for patterns in your results. Predict the numbers of male and female bees in the 9th and 10th generations back. Be ready to explain your reasoning.

Part 2

What if... you examined the ancestry of a species in which both males and females have two parents? How would the patterns be different from those you found in Part 1?

- Work with a partner who is the same sex as you are. Construct your own family tree using blue Snap Cubes to represent males, and purple Snap Cubes to represent females. Here's how:
    - Begin by placing either a purple or blue Snap Cube on your work surface to represent you.
    - Use toothpicks and a cube of each color to represent your mother and father.
    - Continue adding toothpicks and cubes to represent grandparents, great-grandparents, and so on.
    - Making sure to keep all ancestors of the same generation in the same horizontal row, continue the process of adding toothpicks and cubes to trace back 6 complete generations in your family tree.
- Record your data in a chart like the one you made in Part 1.
- Look for patterns in your results. Predict the numbers of male and female ancestors in the 10th generation back. Then generalize your findings by writing algebraic expressions for the numbers of ancestors in the *n*th generation back. Be ready to explain your reasoning.

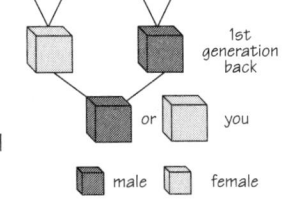

For Your Portfolio

Scientists have found many examples of the Fibonacci sequence occurring in nature. For example, the spiral patterns in pine cones, pineapples, and sunflowers, the leaf arrangements on a branch, and the number of petals on a flower are all related to the Fibonacci sequence. This sequence also appears in architecture and in other fields. Research one or more of these topics and write a short report describing how your topic is related to the Fibonacci sequence.

Napkins and Place Mats

Part 1

Nadia is preparing for a cookout and is setting the picnic table with paper napkins and place mats. She notices that the color scheme on these paper products consists of a solid color in the center surrounded by a contrasting colored border. What relationships can you find among the size of the square napkin, the area of the solid interior, and the area of the contrasting border?

- Work with a partner. Using Color Tiles, make a 3-by-3 model of a square napkin. Use one color (red, for example) to represent the solid interior, and a contrasting color (blue, for example) for the border.
- Build 3 more models of square napkins. Make each of your models one tile longer on each side than in the previous square. Again, use different colors for the interior and for the border.
- As you make each new square, record the dimensions of your model, the dimensions of the interior, the number of tiles in the interior, the number of tiles in the border, and the total number of tiles used. Organize your data in a chart.
- Look for patterns in your data. Predict what tiles would be needed for an 8-by-8 napkin model. Predict what tiles would be needed for a 15-by-15 napkin model. Then generalize the patterns you found by writing algebraic expressions for an *n*-by-*n* napkin model. Add these entries to your chart.
- Be ready to talk about the relationships you discovered.

Part 2

What if... the rectangular place mats, like the napkins, have a solid interior surrounded by a contrasting colored border? Will the numerical relationships formed by this design be the same as those you found for the square napkins?

- Using Color Tiles, make a 3-by-5 model of a rectangular place mat. Use two different colors, one for the interior and another for the border.
- Build 3 more models of place mats, increasing the width by 1 tile and the length by 2 tiles for each subsequent rectangle. Again, use different colors for the interior and for the border.
- As you make each new rectangle, record its dimensions, the dimensions of the interior, the number of tiles in the interior, the number of tiles in the border, and the total number of tiles used. Organize your data in a chart.
- Look for patterns in your data. Predict what tiles would be needed for an 8-by-15 place mat model. Predict what tiles would be needed for a 15-by-29 place mat model. Add these entries to your chart.
- For each place mat model, look to see if you can find a relationship between its width and the rest of its data. Then generalize the relationships you find by writing algebraic expressions to represent the data for a place mat whose width is *n* units.
- Be ready to explain your work.

For Your Portfolio

Explain the methods you used in looking for patterns, analyzing the data, and writing algebraic expressions to generalize your findings.

Pythagoras Delivers the Mail

Part 1

Pythagoras, one of the world's most famous mathematicians, discovered interesting relationships between types of triangles and the lengths of their sides. Can you retrace the steps leading to his remarkable discovery?

- Work with a group of at least 4 students. Each of you should make a different-sized right triangle on your Geoboard.
- Record your triangle on dot paper near the center of the paper.
- Using a ruler, draw a square on each side of your triangle. Make the sides of each square congruent to the side of the triangle on which it is built.
- Find and record the area of each of your three squares. Let the area of one small dot-paper square be the unit of measure.
- Discuss and check the work of the other members of your group.
- Repeat the activity using obtuse triangles and then acute triangles.
- Look for relationships among the areas of the three squares surrounding each type of triangle. Generalize the findings of your group.

Part 2

What if... a letter measuring 1/8" by 11" by 14" is sent to a teacher who has an open-front rectangular mailbox in the school office measuring 5" high by 10" wide by 15" deep. Will the letter fit into the teacher's mailbox without being bent or folded?

- Draw a model of the front of the mailbox on dot paper.
- Discuss with your group how the letter might be positioned in the mailbox, and sketch it on your dot paper drawing.
- Using what you learned in the first activity, decide whether the mailbox can accommodate the teacher's letter. Be ready to justify your conclusions.

For Your Portfolio

Write a paragraph or two describing how you might approach the mailbox problem if the mail was a package instead of a letter.

Polygons, Pegs, and Patterns

Part 1

Did you ever wonder whether the area of a Geoboard polygon might be related to the number of pegs on its boundary and in its interior? What patterns can you find that might prove this to be true?

- Work in a group. Using your Geoboard, explore all possible areas for polygons that contain only 3 boundary pegs. Start by investigating polygons that have 3 boundary pegs and no interior pegs, then 3 boundary pegs and 1 interior peg, 3 boundary pegs and 2 interior pegs, and so on.

- Copy each polygon onto geodot paper and record the number of boundary pegs, the number of interior pegs, and the area.

- Now investigate polygons having 4, 5, and 6 boundary pegs in the same way.

- Organize your findings in a chart. Look for patterns in the data that can be used to help predict the areas of other polygons.

Part 2

What if... you wanted to write a formula for the area of a polygon based upon the numbers of its boundary pegs and interior pegs? How might you do this?

- Look at the data you collected in the first part of the activity. See if you can figure out how the area of a shape could be calculated using the number of boundary pegs (B) and the number of interior pegs (I). Consider simple algebraic expressions containing various operations.

- If you find a formula that seems to work for a particular set of data, test it on other sets. If it doesn't work, try modifying it in some way and then test it again.

- If you have trouble finding a formula that works for all polygons, consider the patterns formed by the data and the types of numbers that appear in the different columns of your charts. Consider the answers to questions such as "When is the area of the polygon a whole number of units? When is it a half number of units?" and "What is the relationship between the area and the number of boundary pegs when there are no interior pegs?"

- Share your ideas with other groups. If you discover a formula that works for polygons with 3, 4, 5, and 6 boundary pegs, test it on Geoboard polygons having 7 or 8 boundary pegs.

- Be ready to share your findings and to talk about the strategies you used in attempting to derive a formula.

For Your Portfolio

Describe the techniques you used in trying to derive your formula. Explain how you went about analyzing the data and formulating ideas to try, and what happened as your investigation proceeded.

The Airline Connection

Part 1

Polygons and their diagonals have many real-world applications, one of which you'll investigate in Part 2 of this activity. But first, can you find a way to predict the number of diagonals that can be made in a polygon?

- Work with a partner. Each of you should make a polygon on the circular side of a Geoboard. Make polygons whose vertices are located at pegs on the circle. Do not use the peg at the center of the circle.
- Pick a vertex and make as many diagonals as you can from that vertex. Repeat this process from each vertex of your polygon.
- Draw your polygon and its diagonals on circular geodot paper. Record the number of sides in your polygon, the number of vertices, the number of diagonals from each vertex, and the total number of diagonals.
- Repeat this process, each time making a polygon with a different number of sides.
- Organize your data with those of your partner and create a chart presenting your findings. Look for patterns in your data.
- Predict the number of diagonals that can be made in a polygon with n sides. Write a formula that can be used to generalize your conclusion.

Part 2

What if... Alpha Airlines has just been granted permission to service 8 new cities in the United States and must hire more pilots? Each pilot will fly direct roundtrip flights between 2 of the 8 cities. If there must be direct flights between every pair of cities, how many pilots must be hired by Alpha Airlines?

- Discuss with your partner how you might use the circular Geoboard and what you learned in Part 1 to solve Alpha's hiring problem. You may want to begin by using your Geoboard to construct a model of the airline's routing plan.
- As you work with your model, you may want to consider the following questions:
    How many flights will originate from one city? How is this problem similar to the problem in Part 1? How is it different? How do the differences affect the problem?
- Using your model, determine the number of pilots that Alpha must hire.
- Write an algebraic expression that could be used to determine the number of pilots who need to be hired if the airline is to service n new cities. Be ready to justify your reasoning.

For Your Portfolio

Write a paragraph or two describing a real-world situation that can be modeled by a polygon and its diagonals. Try to express the situation in terms of a problem that needs to be solved.

Inside Out, Outside In

Part 1

Angela has a hypothesis that the sum of the interior angle measures of a polygon is related to the number of sides in the polygon. Can you help Angela test her theory?

- Working with a partner, find the measure of each angle of each Pattern Block shape. Record your findings by tracing the Pattern Blocks on paper and labeling each angle with its measure. Also record the sum of the angle measures for each shape.

- Make new polygons by combining Pattern Blocks in various ways.

- Copy your polygons onto paper. For each polygon, find the measure of each of its interior angles. Record these measures on your drawings. Also record the number of sides of the polygon, and the sum of its angle measures.

- Organize your findings and look for patterns.

- Use your data to predict the sum of the interior angles of a polygon with 12 sides. Then write an algebraic expression that can be used to find the sum of the interior angle measures of a polygon with *n* sides.

Part 2

What if... you found the measures of the <u>exterior</u> angles of different polygons? Do you think you might find a relationship between the sum of their measures and the number of sides in the polygon?

- For each *convex* shape you recorded in the first activity, use a ruler to extend one of the sides of the polygon at each vertex to form exterior angles.

- Determine and record the measure of each exterior angle of your polygons. Also record the number of sides of each polygon and and the sum of its exterior angle measures.

- Organize your findings and look for patterns.

- Use your data to predict the sum of the measures of the exterior angles of a convex polygon with 12 sides. Then write a general statement about the sum of the exterior angle measures of any convex polygon.

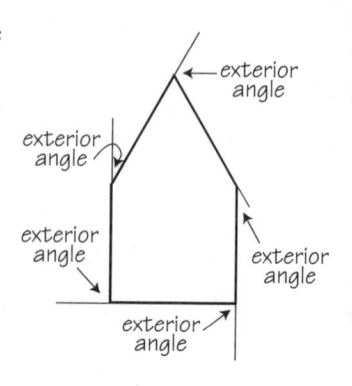

For Your Portfolio

Suppose you know the sum of the interior angles of a particular *regular* polygon (a polygon that is both equilateral and equiangular). Explain how you might figure out the number of sides in the polygon and the measure of each interior and each exterior angle.

Count Square and Countess Triangle

Part 1

Count Square and Countess Triangle enjoy challenging each other with brain teasers. It is the Count's turn to pose one of these mind ticklers to the Countess. He chooses the following problem for her to ponder:
How many squares are there on an 8-by-8 checkerboard?
Can you help the Countess find the solution?

- Working with a partner, use Color Tiles to build an 8-by-8 square to represent the checkerboard.
- Decide on a way to search for all the different-sized squares that are on the checkerboard.
- Record the sizes of the squares you find and the number of squares of each size.
- Look for patterns that might help in your search.
- Determine the total number of squares on the checkerboard. Be ready to tell about how you organized your search.

Part 2

What if... the Countess challenges Count Square with a similar problem based on finding the number of triangles in a large equilateral triangle made from Pattern Blocks? Can you help the Count solve his mind tickler?

- Working with a partner, use green Pattern Block triangles to build an equilateral triangle that has a base equal in length to the sides of 6 green triangles.
- Decide on a way to search for all the different-sized triangles contained in your equilateral triangle.
- Record the lengths of the bases of the triangles you find and the number of triangles of each size.
- Look for patterns that might help in your search.
- Determine the total number of triangles in the large triangle. Be ready to tell about the strategies you used to search for triangles and why you think you've found them all.

For Their Portfolio

Write a letter to the Count and Countess explaining how you would go about finding all the squares on a 20-by-20 Color Tile checkerboard.

Greek Border Designs

Part 1

Almost 2,500 years ago, Greek artisans decorated their pottery with a geometric border called the Greek Key design. Can you predict how many tiles of each color will be needed to build any design in a sequence of Greek Key designs?

- Look at this sequence of Greek Key designs built using red, yellow, and blue Color Tiles.
- Working with a partner, record the number of red, yellow, and blue tiles used in each design. Also record the total number of tiles in each design. Organize your data in a chart.
- Use your Color Tiles to build the next few terms of the sequence and add the information about these terms to your chart. Continue until you see patterns in your data.
- Predict how many tiles of each color will be needed for the 10th design in the sequence (without building it). Do the same for the 100th design.
- Predict how many tiles of each color will be needed for the nth design in the sequence. Express your predictions using algebraic expressions.
- Be ready to talk about how you made your predictions.

Part 2

What if... the number of tiles of each color is limited, and the sequence of designs must contain particular patterns of tile colors? What Greek border designs can be created?

- Study the chart at right.
- Working with your partner, create a sequence of designs that fits the data in the chart. To do this, first determine the number of red, yellow, and blue tiles that are added to each design to produce the next design in the sequence. These tiles make up what is called the "repeat" portion of the design. As its name suggests, the repeat will appear over and over again in your designs.
- Create a design using the tiles in the repeat. Add on any additional tiles needed to satisfy the data for design 1.

design number	number of red tiles	number of yellow tiles	number of blue tiles	total number of tiles
1	6	4	3	13
2	7	8	6	21
3	8	12	9	29
4	9	16	12	37
5	10	20	15	45
6	11	24	18	53

- Make two copies of your repeat and link them together end to end. Add on any additional tiles indicated by the data for design 2. Try to maintain pattern and color symmetry at both ends of your border design.
- Build three more terms in your sequence. Record and color your design on Color Tile grid paper.
- Create several other Greek border designs using the same data. Record your designs on grid paper and compare them.

For Your Portfolio: Create you own Greek border design using Color Tiles. Record and color it on Color Tile grid paper. Then write a paragraph or two describing the underlying patterns in your design. Tell how the numbers of tiles needed for longer versions of your border design can be calculated.

Table for 63, Please

Part 1

> Matt and Jamal own a company that rents tables and chairs for parties. They have 100 tables of each shape in their inventory. These tables can be joined together to provide seating accommodations for larger groups of people. Can you predict how many people can sit at one large table made by joining 100 tables?
>
> - Work with a partner. Use a green Pattern Block triangle to represent one table. Figure that one person can be seated on each side of a triangular table.
> - Build larger tables by joining triangles side to side. Add tables in such a way so that each new table added shares only one side with the existing table. Record how many people can sit at each larger table. Make a chart to keep track of your findings.
> - Continue until you have built at least 6 or 7 tables of different sizes.
> - Look for patterns in the numbers you recorded. Use these patterns to predict how many people can sit at a table built from 100 triangular tables. Then generalize your findings and predict how many can sit at n triangular tables.
> - Repeat the process of building larger tables, first using orange Pattern Block squares as tables, and then using yellow Pattern Block hexagons. Be sure that each new shape added shares only one side with the existing table.
> - For each shape, look for patterns that would help you predict how many people can sit at a table built from 100 such tables, and then generalize your findings to predict how many can sit at n tables.

Part 2

> **What if...** Matt and Jamal are asked to provide tables and chairs for a party Rosie is having for 63 people? Using any combination of triangular, square, and hexagonal tables, how can they provide seating for exactly 63 people?
>
> - Consider seating patterns that might be generated by combining a variety of different-shaped tables. Use the materials and information you gathered from the first activity.
> - Look for table combinations that can accommodate exactly 63 people. Build or sketch models of your arrangements.
> - Decide whether any of the table combinations would be more desirable than others. Be ready to explain your reasoning.

For Your Portfolio

Write a letter to Matt and Jamal describing the combination of triangular, square, and hexagonal tables you think they should supply for Rosie's party of 63 people and why you think it is the best choice. Include any diagrams or instructions that might be helpful.

Birthday Cakes

Part 1

Anastasia is having a birthday party and is preparing to slice her round birthday cake. She wonders if there is a way to predict the number of pieces of cake that can be obtained by cutting the cake a certain way. Can you help Anastasia determine how many people she can serve?

- Using the Circular Geoboard to represent Anastasia's birthday cake, join any two points on the circle with a rubber band representing the first cut. (Do not use the center point on the Geoboard as one of the two points.)
- Draw a diagram of the cake and the first cut on circular geodot paper. Record the number of points connected, the total number of cuts made (in this case, 1), and the number of servings of cake obtained.
- Choose a third point on your Geoboard cake and make a series of cuts using rubber bands that join your new point to the points you used before. Select this new point in a way that maximizes the number of pieces of cake formed when the cuts are made. In other words, try to get as many pieces as you can.
- Record your cake and data as you did before.
- Repeat the above procedure, adding a 4th point and then a 5th point. Be sure to record a diagram and the data for each point added.
- Organize your data and look for patterns that can help you predict the maximum number of servings of cake that will be formed by adding a 6th point. Then test your prediction by adding a 6th point, connecting it to all the others, and counting the number of pieces of cake obtained.
- Share and compare your work with your partner. Be ready to discuss your findings.

Part 2

What if... Anastasia thinks that at her next birthday party it would be easier for her to serve a rectangular cake? Would she find that the relationship between the number of cuts and the number of slices is different from what she found with her circular cake?

- On your Geoboard, use a rubber band to mark off a rectangle whose length is 1 unit longer than its width. Let this rectangle represent Anastasia's birthday cake. Use rubber bands to "slice" the cake into 1-by-1 squares.
- Copy your cake onto geodot paper. Record its dimensions, the number of cuts made, and the total number of servings of cake obtained.
- Make several different Geoboard cakes, each with length 1 unit longer than the width. Cut each cake into single square servings as described. Make a drawing of your cake and record the data as you did with your first cake.
- Organize your data in a chart and look for patterns.
- Generalize your findings by writing algebraic expressions that can be used to predict the number of cuts and the number of pieces of cake for rectangular cakes with width n units and length 1 unit longer. Use dot paper to test your hypotheses on other rectangular cakes whose lengths are 1 unit longer than their widths. Be ready to discuss your findings.

For Your Portfolio

Describe how the two parts of this activity were alike and how they were different. Discuss the roles of patterns and reasoning. Include any conclusions you can make about using patterns to make predictions.

Visual Effects

Part 1

Although mathematical and artistic abilities are located in different hemispheres of the brain, you can strengthen both of these abilities by working with visual design sequences. Can you extend the patterns for the two sequences below?

- Study the first 4 terms of Sequence #1.
- Working with a partner, use Cuisenaire Rods to build the next six terms of Sequence #1.
- Draw and color the first ten terms of Sequence #1 on 1-centimeter grid paper.
- A set of rules for building Sequence #2 is given below. Using Cuisenaire Rods and Tangram pieces, work with your partner to build the first eight terms of Sequence #2.

Sequence #2

Every term contains either a small or a medium Tangram triangle. The small triangle appears in every other term beginning with the first. It sits on one of its shorter sides with its right angle in the lower right corner. The medium triangle, whenever it appears, balances on the vertex at its largest angle. To the right of each triangle is a Cuisenaire Rod positioned horizontally. The rods increase in length by 1 unit from term to term, beginning with a rod whose length is 1 unit. For odd-numbered terms, the rod is attached to the side of the triangle with bottom edges in alignment. For even-numbered terms, the rod is attached at the top, vertex to vertex, with top edges of both blocks in alignment.

- Draw and color the first eight terms of Sequence #2 by tracing the Cuisenaire Rods and Tangram pieces on plain paper.
- Compare your drawings with those of another group. If they differ, discuss and resolve your differences.

Part 2

What if... you and your partner were asked to design a sequence using both Cuisenaire Rods and Tangram pieces? What would your sequence be like?

- Work so that other groups cannot see the sequence you are designing.
- Using both Cuisenaire Rods and Tangram pieces, design and build the first six terms of an original sequence. The patterns in your sequence can be based on color, shape, position, number of pieces, or a combination of these and other characteristics.
- Draw the six terms of your sequence on plain paper. This will be the answer key.
- On another sheet of paper, write a set of rules that can be used to generate the terms of your sequence. Be as mathematically concise as possible.
- Exchange your rules with those of another group. Try to build the first six terms of the sequence described by the other group's rules.
- When both groups have finished, check the sequences against the answer keys. Discuss any differences until agreement is reached.

For Your Portfolio

Think about the kinds of rules you wrote for your sequence. Then give an example of a rule that, although it is mathematically accurate, could be interpreted to generate two (or more) different sequences. Discuss the possible interpretations of your rule and ways to make it more precise.

Pascal Pastimes

Part 1

Eugene loves to play pinball and believes there is a way to figure out how many paths the silver ball could follow as it wends its way down through the bumpers. Can you help Eugene explore a method for finding the number of different paths the ball could take?

- Working with a partner, use the Geoboard to represent the pinball machine, its pegs to represent the bumpers, and rubber bands to represent the path of the ball. Position the Geoboard with a corner at the top and use lettered sticky dots to identify bumpers.
- Find all the paths the pinball could take as it travels from bumper A to bumper B, from A to C, from A to D, and so on, according to the following rule: As the ball moves from one peg (or bumper) to the next, it must travel downward to one of the two pegs closest to it in the next row.
- Look for all possible paths to each point. Keep track of the paths you find. (Geodot paper may be helpful.) Look for patterns that may help you predict the number of paths to particular pegs.
- On a clean sheet of geodot paper, record the *number of paths* you find from A to each bumper.
- Use your patterns to predict the number of different paths from A to Y without actually making them on your Geoboard. Be ready to explain your reasoning.

Part 2

What if... Eugene is also curious about a game where pennies are tossed a certain number of times and the numbers of heads and tails thrown determines the winner? He might wonder if the outcomes follow a predictable pattern. What can you help Eugene discover about this game?

- Working with your partner, place sticky dots on both sides of each of 4 Color Tiles. Let the tiles represent pennies. On each "penny," mark an "H" (heads) on one side and a "T" (tails) on the other.
- Have one partner act as recorder while the other partner tosses the tiles.
- Toss all 4 marked Color Tiles at the same time onto a flat surface. Count the number of "heads" and "tails" and record the outcome with a tally mark in a chart like the one shown.
- Toss the tiles 15 more times, recording each outcome in the chart.
- Calculate the totals and look for patterns.
- Conduct similar experiments, tossing 5 marked Color Tiles 32 times and then 6 marked Color Tiles 64 times. Record the outcomes in charts like the one to the right.

Heads:	4	3	2	1	0
Tails:	0	1	2	3	4
Tally:					
Totals:					

- Discuss your results with your partner. What patterns do you see? What predictions can you make? What relationships do you see between the outcomes of this problem and the pinball path problem?

For Your Portfolio

Investigate other applications of Pascal's Triangle and write a few paragraphs describing one application that you find particularly interesting.

Beehive Buzz

Part 1

A beehive is abuzz with activity as its bee colony, consisting of queen bees, drones, and workers, moves in to make honey in a 7-cell honeycomb. If different-shaped Pattern Blocks are used to represent the honeycomb and the three types of bees, how many different combinations of bees can completely occupy the honeycomb?

- Work with a partner. One of you should work on Honeycomb A while the other works on Honeycomb B.
- Each of you should begin by placing 7 yellow hexagon blocks together to make the honeycomb shown at the right.

Honeycomb A
- Your honeycomb will be occupied by queen bees (red trapezoids) and workers (green triangles).
- Use only red and green Pattern Blocks to cover your hexagon honeycomb.
- Record the number of red queens, the number of green workers, and the total number of bees in your honeycomb.
- Now use different combinations of your Pattern Block bees to cover your hexagon honeycomb. See how many different combinations you can find. Organize your findings in a table.

Honeycomb B
- Your honeycomb will be occupied by drones (blue rhombuses) and workers (green triangles).
- Use only blue and green Pattern Blocks to cover your hexagon honeycomb.
- Record the number of blue drones, the number of green workers, and the total number of bees in your honeycomb.
- Now use different combinations of your Pattern Block bees to cover your hexagon honeycomb. See how many different combinations you can find. Organize your findings in a table.

- Share and compare your findings with those of your partner. Look for relationships among the numbers and types of bees occupying the honeycombs.

Part 2

What if... the 7-cell honeycomb is occupied by queens, drones, and workers? What combinations of bees might be found in the honeycomb?

- Working with your partner, build one 7-cell honeycomb using yellow hexagons.
- Using red trapezoids (queen bees), blue rhombuses (drones), and green triangles (workers), find a way to cover your hexagon honeycomb.
- Record the number of red queens, blue drones, and green workers, and the total number of bees in your honeycomb.
- Now use different combinations of Pattern Block bees to cover your hexagon honeycomb. See how many different combinations you can find. Organize your findings in a table.
- Look for relationships among the numbers and types of bees occupying the honeycomb.
- Predict the maximum number of queens, drones, and workers that could occupy a honeycomb with n cells. Be ready to justify your conclusions.

For Your Portfolio: Write an explanation of the strategies you used to organize your data as you searched for solutions to the honeycomb problems.

Carol's Kite Kits

Part 1

Carol designs kite kits of various sizes and styles. One of the most popular styles is the box kite. The frame is constructed of lightweight wooden strips with brightly colored bands of paper attached to the top and bottom of the frame. Can you help Carol determine the amounts of wood and paper needed for different-sized box kite kits?

- Working with a partner, use Snap Cubes to build a model of kite #1 with dimensions 1 x 1 x 2. Draw a model of this kite on isometric dot paper.
- Determine and record the total length of wooden strips needed to build the frame of kite #1. Let the edge of a cube represent 1 unit.
- Two bands of paper are needed to finish the kite's construction. Each band is one fourth the height of the frame. Determine and record the total amount of paper (in square units) needed for kite #1.
- Build kite models with dimensions 2 x 2 x 4, 3 x 3 x 6, and so on, where the height is twice the side length of the square base. Draw each model on dot paper and record the amounts of wood and paper needed to build it.
- Organize your data in a chart. Look for patterns to help you complete the data for the first six kite kits. Then predict the amounts of materials needed to build a kite measuring 10 x 10 x 20.
- Write algebraic expressions that generalize the patterns you find. Be ready to explain your findings.

Part 2

What if... Carol decides to make kits for diamond-shaped kites? If the frame of each kite is made from two wooden strips and is covered with a thin layer of paper, how much wood and paper would be needed for each kit?

- Working with a partner, use Snap Cubes to build a model of the frame for the first kite as follows:
  - Start with one cube representing the portion of the frame where the wooden strips overlap.
  - Attach one cube to the top face of the cube, one to its left face, one to its right face, and two to its bottom face.
- Draw your model of this kite frame on grid paper and label it kite #1. Determine and record the lengths of each of the two pieces of wood needed to build the frame, as well as the total amount of wood needed. Let the edge of a cube represent 1 unit.
- Determine and record the total amount of paper (in square units) needed to cover the frame of kite #1.
- Build models of other diamond-shaped kite frames where the number of cubes attached to the bottom face of the overlap cube is twice the number of cubes attached to the top, left, and right faces. Draw and number each model on grid paper and record the amounts of wood and paper needed to build each kite.
- Organize your data in a chart. Look for patterns.
- Write algebraic expressions that generalize the patterns you find. Be ready to explain your findings.

For Your Portfolio

Design your own kite. Draw a picture of it on isometric dot paper, grid paper, or plain paper. Then write a letter to Carol, asking her to consider making and selling kits for your kite. Describe how your kite is constructed, and explain ways to determine the amounts of materials needed for kits of different sizes.

Geodot Paper

Patterns/Functions ◆ Grades 7-8

Circular Geodot Paper

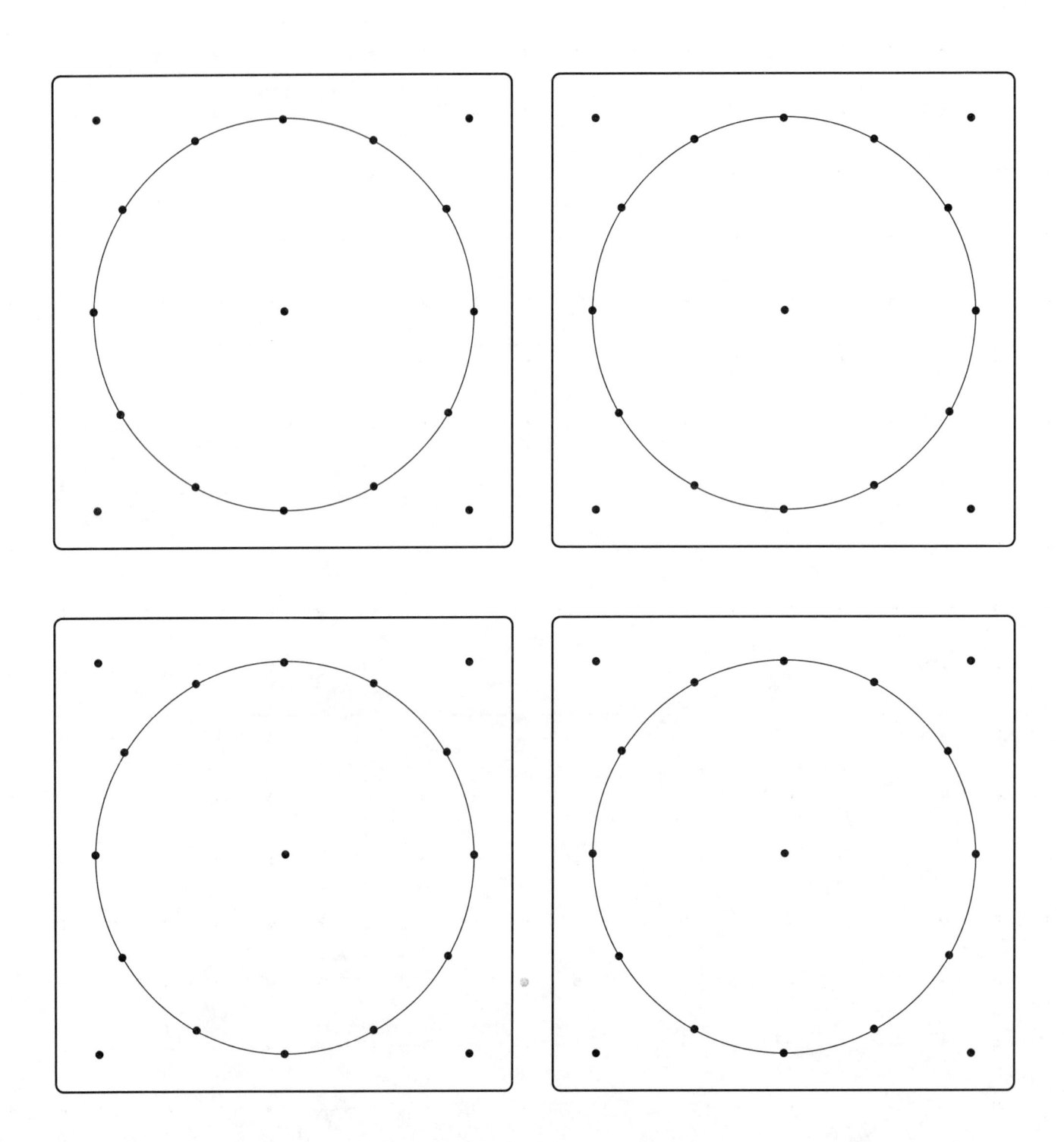

118 the Super Source™ ♦ Patterns/Functions ♦ Grades 7-8 ©Cuisenaire Company of America, Inc.

Dot Paper

Color Tile Grid Paper

1-Centimeter Grid Paper

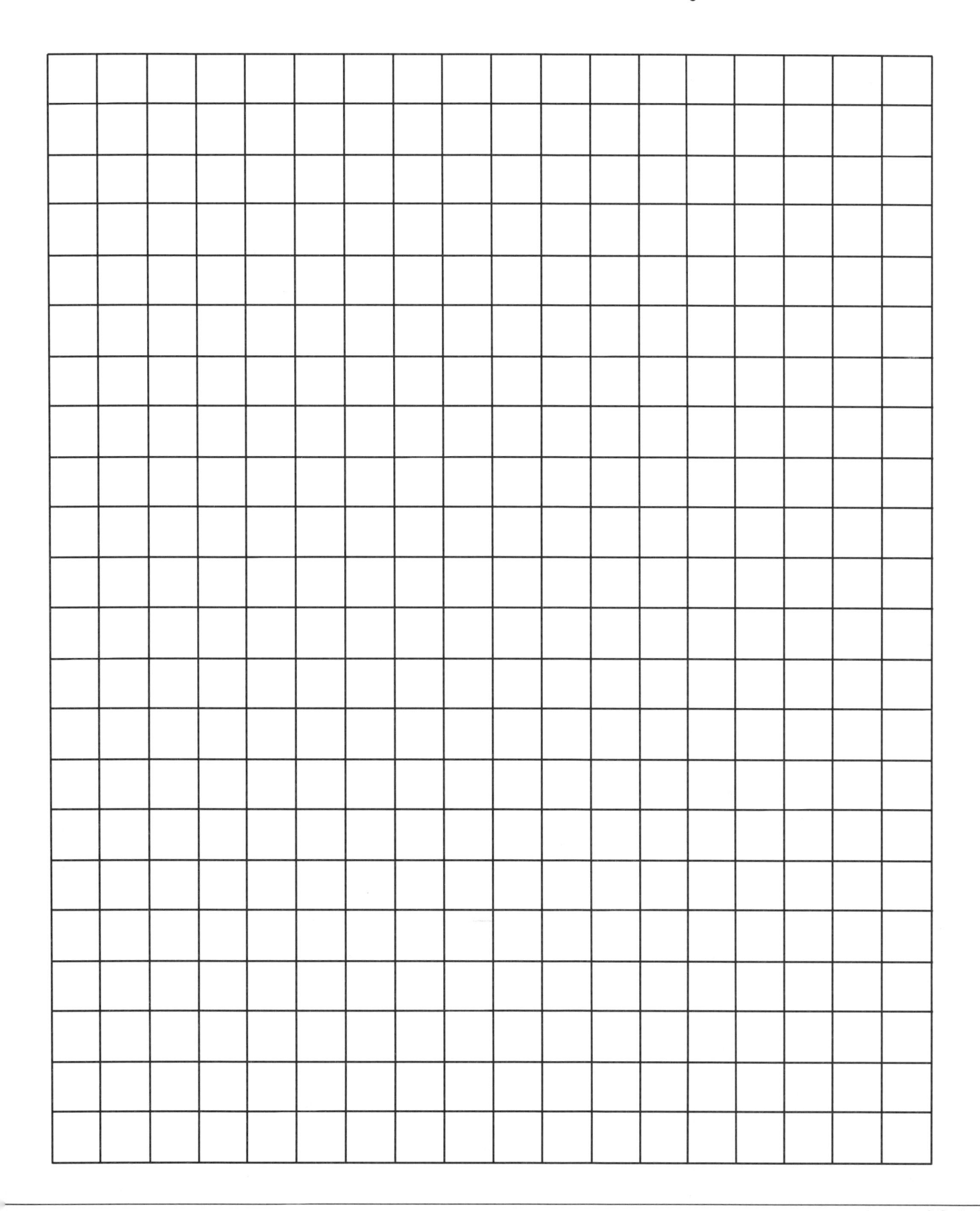

Pattern Block Triangle Paper

Isometric Dot Paper